# SpringerBriefs in Computer Science

*Series Editors*
Stan Zdonik
Peng Ning
Shashi Shekhar
Jonathan Katz
Xindong Wu
Lakhmi C. Jain
David Padua
Xuemin Shen
Borko Furht
VS Subrahmanian

For further volumes:
http://www.springer.com/series/10028

SpringerBriefs in Computer Science

Eric Rosenberg

# A Primer of
# Multicast Routing

 Springer

Eric Rosenberg
AT&T Labs
Middletown, NJ, USA
ericr@att.com

ISSN 2191-5768      e-ISSN 2191-5776
ISBN 978-1-4614-1872-6      e-ISBN 978-1-4614-1873-3
DOI 10.1007/978-1-4614-1873-3
Springer New York Dordrecht Heidelberg London

Library of Congress Control Number: 2011946218

Printed on acid-free paper

Springer is part of Springer Science+Business Media (www.springer.com)

# Preface

This is an introduction to multicast routing, which is the study of methods for routing from one source to many destinations, or from many sources to many destinations. Multicast is increasingly important in telecommunications, for such applications as software distribution, video on demand, and file transfers.

The intended audience of this primer includes

- telecommunication network designers and architects,
- researchers in telecommunications and optimization,
- upper class undergraduate students, and graduate students, studying computer science, electrical engineering, or operations research.

We assume the reader is familiar with computing a shortest path in a network, e.g, by Dijkstra's method. No prior knowledge of telecommunications or routing protocols is assumed, although it will undoubtedly make the reading easier. Although a few of the mathematical results are quite deep, they are presented without proof (but with references to the literature), and can be skipped with no loss of understanding of subsequent sections.

Both obsolete and currently used methods are examined, in order to prevent researchers from reinventing previously proposed methods, and to provide insight into why some methods did become obsolete. The reference list has over 100 entries. It would not be difficult to find another 100 papers worthy of review or mention. In particular, several areas of multicast are quite active, including wireless multicast, label switched multicast, and aggregated multicast. The selection of papers for this primer naturally reflects both space considerations and the author's own interests. In addition to the usual journals and conference proceedings, the IETF web site is a rich source of multicast protocol standards and drafts. The web sites of major router vendors are also valuable sources for both protocol descriptions as well as implementation guidelines and configuration details.

The challenges faced by a small company wishing to implement multicast are typically quite different from the challenges faced by a global service provider. One size does not necessarily fit all when it comes to multicast. It is hoped that this primer will guide both academic researchers and engineers responsible for designing and implementing multicast networks.

My sincere thanks to Lee Breslau, Josh Fleishman, Apoorva Karan, Thomas Kernen, Ron Levine, Bob Murray, Han Nguyen, Marco Rodrigues, Samir Saad, Steve Simlo, and especially to Yiqun Cai, Ken Dubose, Don Heidrich, Maria Napierala, Eric Rosen, and Mudassir Tufail, for stimulating discussions and comments on drafts of this primer. Their suggestions have been enormously helpful.

Middletown, NJ                                                          *Eric Rosenberg*
September, 2011

# About the Author

Eric Rosenberg received a B.A. in Mathematics from Oberlin College and a Ph.D. in Operations Research from Stanford University. He works at AT&T Labs in Middletown, New Jersey (email: ericr@att.com). Dr. Rosenberg has taught undergraduate and graduate courses in optimization at Princeton University and New Jersey Institute of Technology. He holds seven patents and has published in the areas of convex analysis and nonlinearly constrained optimization, computer aided design of integrated circuits and printed wire boards, and telecommunications network design and routing.

# Contents

# Acronyms

Following the definition is the chapter/section the term is first introduced.

AD        auto-discovery 7.3
AMT       automatic IP multicast without explicit tunnels 4.4
AS        autonomous system 1.1
ASM       any source multicast 1.1
BGP       border gateway protocol 1
CAN       content addressable network 4.3
CBT       core based tree 3.6
CE        customer edge router 7.1
DF        designated forwarder 3.7.3
DVMRP     distance vector multicast routing protocol 3.4
FIB       forwarding information base 3.10
GRE       generic routing encapsulation 1.4
IETF      Internet engineering task force 1.1
IGMP      Internet group management protocol 3.1
IGP       interior gateway protocol 3.8.4
IP        Internet Protocol 1.2
I-PMSI    inclusive provider multicast service instance 7.4
ISP       Internet service provider 1.1
LAN       local area network 1.1
LFIB      label forwarding information base 3.10
LSA       link state advertisement 3.5
LSP       label switched path 3.10
LSR       label switched router 3.10
MDT       multicast distribution tree 7.2
MFIB      multicast forwarding information base 6.4
MLD       multicast listener discovery 3.7.4
MOSPF     multicast open shortest path first 3.5
MP2MP     multipoint-to-multipoint 3.10
MP-BGP    multiprotocol border gateway protocol 5.1
MPLS      multiprotocol label switching 3.10
MSDP      multicast source distribution protocol 3.8.4
MST       minimal spanning tree 1.4
MTI       multicast tunnel interface 7.2.1
MVPN      multicast virtual private network 7
OIL       outgoing interface list 2.2
OSPF      open shortest path first 1
P2MP      point-to-multipoint 3.10

PE        provider edge router 7.1
PIM       protocol independent multicast 2.3
PIM-SM    PIM-Sparse Mode 3.7.2
PMSI      provider multicast service instance 7.3
RD        route distinguisher 7.1
RP        rendezvous point 2.3
RPF       reverse path forwarding 1.5.1
RT        route target 7.1
SA        source active 5.2
SP        service provider 7
S-PMSI    selective provider multicast service instance 7.4
SPT       shortest path tree 3.7.2
SSM       source specific multicast 1.1
TTL       time to live 3.1
VPN       virtual private network 7
VRF       virtual routing and forwarding table 7.1

# Chapter 1
# What is Multicast Routing?

Consider a telecommunications network consisting of a set of nodes connected by arcs. The network might be, e.g., the Internet, or the private network offered by a service provider such as AT&T. A node represents a physical device, such as a switch or router, connected to other devices. The term *switch* is typically used to refer to a device performing layer 2 *data link* functionality (e.g., Asynchronous Transfer Mode (ATM) or Frame Relay) in the OSI model [102], while a *router* performs layer 3 *network* functionality in the OSI model. For brevity, by *node* we mean either a switch or a router. An arc represents a communications pathway, such as a fiber optic cable or a radio (wireless) link. Suppose a given source node in the network wishes to send the same data stream, composed of packets, to one or more destination nodes in the network. The data stream might be, e.g., an all-employee broadcast, software distribution, a file transfer, financial information, video on demand, or emergency management communications.

We want to determine the path that should be used to send the stream to each destination node. When there is a single destination node, this problem is known as *unicast routing*, and is typically solved by computing a shortest path between the source and the destination. The path might be computed by a *link state* method such as ISIS [52], *Open Shortest Path First* (OSPF) [77], or *Private Network-Network Interface* (PNNI) [8], or by a *distance vector* method such as RIP [68]. (In link state methods, each node has its own view of the network topology, and computes routes to all destinations; the classic link-state method is Dijkstra's method [5]. In distance vector methods, nodes exchange and update routing tables; the classic distance vector method is the Bellman-Ford method [5].) The source node forwards each packet to the next node on the shortest path, this second node forwards the packet to the third node, and this hop-by-hop forwarding terminates when the packet is received by the destination node (a *hop* is a synonym for an arc). Unicast routing is "easy," since the complexity of computing a shortest path in a network with $N$ nodes and $A$ arcs is, e.g., $O(N^2)$ for Dijkstra's method for dense networks, and $O(A + N log N)$ for a Fibonacci heap implementation [5] (a function $f(x)$ is said to be $O(g(x))$ if there is a positive constant $c$ such that $|f(x)| \leq c |g(x)|$ for all $x \geq 0$).

When the stream is required to be transmitted to every other node in the network, the problem is known as *broadcast routing* or simply *broadcasting*. Broadcasting is used, e.g., in ad-hoc mobile networks, where node mobility causes frequent link failures [56]. When the stream is required to be transmitted to only a specified subset of nodes in the network, the problem is called *multicast routing* or simply *multicasting*. Thus unicast is the special cast of multicast where there is one destination, and broadcast is the special case where all nodes are destinations.

## 1.1 Groups, Sources, and Receivers

In multicast routing, the basic constructs are sources, receivers, and groups. A source is an end user that originates a data stream. A receiver is an end user wishing to receive a data stream. Each source is locally connected to (i.e., *subtends*) a nearby node, typically either by a direct connection or by an Ethernet *Local Area Network* (LAN). A source might also be connected to a second nearby node, in case the connection to the first node, or the first node itself, fails. Similarly, each receiver subtends a nearby node, typically by a direct connection or Ethernet LAN, and might also be connected to a second nearby node. We refer generically to an end user source or receiver as a *host*. There may be no hosts subtending some nodes; such nodes are called *via* nodes.

A *multicast group* is a set of receivers with a common interest. Note that this definition makes no mention of a source. For example, if the group is the set of students registered for an online course offered by a university, the receivers are the computers of the online students, and the source might be the router connected to the video camera in the classroom where the instructor is teaching, A second example is a financial institution such as a brokerage house, where the group is the set of stock traders, distributed around the globe, who communicate with each other about stock offerings. The set of receivers is the entire set of computers used by the traders; since each trader typically needs to send financial data to other traders, the set of sources is also the entire set of computers used by the traders.

In the first example, there is a single source, and the set of receivers is relatively static for the duration of the online lecture, since students must pre-register for the online course (although some students might sign-in late or sign-off early). In the second example, both the set of sources and the set of receivers are dynamic, depending on which traders are working that day, or are interested in some stock offer. The financial institution might have multiple multicast groups, e.g., one for bond traders and one for stock traders.

In the *any source multicast* (ASM) model [31], there is no limit on the number of multicast groups, or on the number of sources and receivers for a group. (While the ASM model assumes no limits, in practice equipment vendors may impose limits due to, e.g., processing or memory limitations.) Group membership is dynamic; receivers can join or leave a multicast group at any time. Similarly, the set of sources for a group can vary over time.

Sources and receivers can be physically located anywhere in the network. Any host can be a source or receiver for a multicast group. A given receiver can simultaneously join multiple groups, and a given source can simultaneously send to multiple groups. A source is not required to know the location or identities of receivers, and a source for a group need not join the group.

A refinement of the ASM model, called *source specific multicast* (SSM) and studied in Section 3.7.4, allows, for a given group $g$, a receiver to select the specific sources for this $g$ from which it will receive a stream. SSM prevents a receiver from being deluged with streams from all sources for a given group, and the paths generated with SSM have delay no larger, and may be smaller, than the paths generated with ASM. We assume the ASM model throughout this primer, except where SSM is specifically discussed.

Typically, in a small geographic area like a university campus, the hosts are interconnected by a *multi access subnet* (e.g., a LAN), and the subnet also connects one or more routers. One of the routers on the subnet will be the primary router, and connections from the subnet to the outside world are via the primary router. Usually a secondary router is configured to take over if the primary router fails. We say that the active router is the *local router* for any source or receiver host on the subnet.

The Internet contains a very large number of *Autonomous Systems* (ASs), where an AS is a set of one or more networks administered by the same organization, e.g., a government, corporation, or university. A given AS is administered by only one organization. However, a given organization, e.g., an *Internet Service Provider* (ISP), might administer multiple ASs. We also use the term *domain* to refer to an AS. The *Internet Engineering Task Force* (IETF) is the organization that issues standards documents that specify the algorithms and procedures used for routing in the Internet and in Virtual Private Networks (Chapter 7).

## 1.2 Addressing

Although Layer 2 protocols such as Frame Relay and ATM can support multicast, currently the vast majority of multicast is over Layer 3 networks running the *Internet Protocol* (IP). For IP Version 4 (IPv4) networks, a multicast group is identified by a 32 bit IP address confined to a specific range. The 32 bits in each IPv4 address are divided into 4 octets of 8 bits each, so each octet can express a (decimal) number from 0 to 255. Address assignment is controlled by the *Internet Assigned Numbers Authority* (IANA) [50].

An *address prefix* $A/N$ consists of an IP address $A$ and an integer *mask* $N$, and represents the set of IP addresses whose leftmost (i.e., most significant) $N$ bits match the first $N$ bits in the binary expansion of $A$. For example, consider the address prefix 209.12.0.0/15. Since the address has the binary expansion

11010001.00001100.00000000.00000000, the address prefix represents any IP
address of the form 11010001.0000110x.xxxxxxxx.xxxxxxxx, where "x" is the
wild card (either 0 or 1). Thus this IP prefix represents IP addresses in the
range 209.12.0.0 to 209.13.255.255. This is illustrated in Table 1.1, where byte
1 is the leftmost byte.

|         | byte 1   | byte 2   | byte 3   | byte 4   |
|---------|----------|----------|----------|----------|
| Address | 11010001 | 00001100 | 00000000 | 00000000 |
| Mask    | 11111111 | 11111110 | 00000000 | 00000000 |
| Prefix  | 11010001 | 0000110x | xxxxxxxx | xxxxxxxx |

**Table 1.1** Address prefix

For IPv4, the first (i.e., leftmost) 4 bits of the first octet of a multicast
group address must be 1110, so the multicast address range is 224.0.0.0 to
239.255.255.255. An address in this range is known as a *class D* address [77].
Some class D addresses are reserved for special purposes [23]. For example,
the group address 224.0.0.5 sends to all routers in an OSPF domain. Addresses in the range 224.0.1.0 to 238.255.255.255 are called *globally scoped*
addresses, and are used when the multicast groups span various organiza-
tions and the Internet. Addresses in the range 232.0.0.0/8 are reserved for
Source Specific Multicast. Thus the first 8 bits of any address in this range
must be (decimal) 232, and the remaining 24 bits can be anything. Addresses
in the range 239.0.0.0/8 are called *limited scope* or *administratively scoped*
addresses, and are used when the multicast group has a limited span, such as
a single university or company [24]. An organization using administratively
scoped addresses typically configures its routers to prevent traffic (i.e., a data
stream) with this group address from being sent outside the organization. In
general, a group address does not convey any geographic information, since
the group sources and receivers may be scattered around the globe.

A host wishing to send to multicast group $g$ uses the IP address of $g$ as
the destination address. When a source sends a stream to a multicast group,
the source address of each packet in the stream is the unicast address of the
source. For IPv4 networks, the source address is also a 32 bit address, which
must lie outside of the range reserved for multicast addresses. Since the pool
of available IPv4 addresses is rapidly dwindling, 128 bit IP Version 6 (IPv6)
addresses have been introduced. For IPv6 multicast addresses, the first octet
must be 0xFF (i.e., binary 11111111). Any other value for this octet identifies
the address as a unicast address. IPv6 multicast addresses are defined in [47],
and the rules for assigning new IPv6 multicast addresses are defined in [44].

*GLOP* [70] is a multicast addressing method for IPv4 networks. Curiously,
GLOP is not an acronym; it does not stand for anything. The term was simply
created to refer to this addressing method. For a four octet IPv4 address,
GLOP sets the first octet to 233, sets the second and third octets to the 16
bit Autonomous System number [102], and the last octet is locally assigned.

## 1.3 The Efficiency of Multicast

Suppose a given source node $s$ wishes to send a stream of bandwidth $b$ to a set of $k$ other nodes in a network. One way to accomplish this is to have $s$ create $k$ copies of the stream, and send each of the copies to its destination using unicast. This approach, known as *unicast replication*, is particularly efficient when $b$ and $k$ are small; in this case the extra node overhead and bandwidth required for replication may more than compensate for not having to implement a multicast routing protocol. Similarly, if the streams are very bursty (frequent periods of inactivity between packets), then the overhead of repeatedly building and tearing down trees argues for replication.

Alternatively, we can create a tree, rooted at $s$, connected to each of the $k$ destinations. Replication of the data stream occurs only at branching points of the tree. For example, consider Figure 1.1, which shows a network of 10 nodes interconnected by 9 arcs. Suppose the source node $s$ is node 1, which

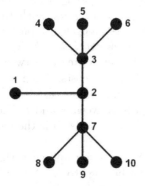

**Fig. 1.1** Multicast vs. unicast

must send a data stream of bandwidth $b$ to nodes 4, 5, 6, 8, 9, 10. With unicast routing, node 1 would need to send 6 copies of the data stream, so the total bandwidth on arc $(1, 2)$ would be $6b$; the total bandwidth on arcs $(2, 3)$ and $(2, 7)$ would be $3b$, and the total bandwidth on each of the other arcs would be $b$. With multicast, node 1 sends a single copy of the stream to node 2, which creates 2 copies of the stream. One copy goes to node 3, which creates 3 copies (for nodes 4, 5, 6), and one copy goes to node 7, which creates 3 copies (for nodes 8, 9, 10). This illustrates the bandwidth savings in the network with multicast. An additional advantage of a tree over unicast replication is that, with unicast replication, node 1 must manage 6 outgoing streams, rather than 1 outgoing stream with a tree. Similarly, node 2 must manage 6 outgoing streams with the unicast approach; with a tree, node 2 manages only 2 outgoing streams. Since incremental memory and processing is required for each stream managed by a node, a tree may lessen the burden on nodes relative to unicast replication.

## 1.4 Mathematical Formulation

Consider a directed network $(\mathcal{N}, \mathcal{A})$, where $\mathcal{N}$ is the set of nodes and $\mathcal{A}$ is the set of arcs. We associate with arc $a \in \mathcal{A}$ a non-negative cost $c_a$. We also denote by $(i, j)$ the directed arc from node $i$ to node $j$; the cost of this arc is $c_{ij}$. A directed network formulation allows for the general case of asymmetric arc costs (i.e., $c_{ij} \neq c_{ji}$); if all costs are symmetric, then an undirected network formulation is appropriate. We use the terms *network* and *graph* interchangeably. Two nodes are *adjacent* or *neighbors* if they are connected by a single arc. A node $n \in \mathcal{N}$ is a *leaf* node if exactly one arc is incident to (i.e., touches) $n$.

A *path* between two nodes $s$ and $t$ is an ordered sequence of $p$ arcs $(s, n_1), (n_1, n_2), \cdots, (n_{p-2}, n_{p-1}), (n_{p-1}, n_p)$ such that $n_p = t$. If $s = t$ and $p \geq 2$ then the path is a *cycle*; i.e., a cycle is a path of two or more arcs forming a loop. The graph $(\mathcal{N}, \mathcal{A})$ is *connected* if there is a path between each pair of nodes; it is *complete* if there is an arc between each pair of nodes. The *cost* of a path is the sum of the arc costs, taken over all arcs in the path. A path between nodes $s$ and $t$ is a *shortest path* if its cost is less than or equal to the cost of any other path between $s$ and $t$. The *diameter* of $(\mathcal{N}, \mathcal{A})$ is $\max\{c^\star(i, j) \mid i \in \mathcal{N}, j \in \mathcal{N}\}$, where $c^\star(i, j)$ is the length of the shortest path between nodes $i$ and $j$.

A *subgraph* $(\bar{\mathcal{N}}, \bar{\mathcal{A}})$ of $(\mathcal{N}, \mathcal{A})$ is a graph such that $\bar{\mathcal{N}} \subseteq \mathcal{N}$, $\bar{\mathcal{A}} \subseteq \mathcal{A}$, and each endpoint of an arc in $\bar{\mathcal{A}}$ belongs to $\bar{\mathcal{N}}$. A *tree* in $(\mathcal{N}, \mathcal{A})$ is a connected subgraph with no cycles. The cost of a tree is the sum of the arc costs over all arcs in the tree. A *spanning tree* of a connected graph $(\mathcal{N}, \mathcal{A})$ is a tree $(\mathcal{N}, \bar{\mathcal{A}})$ where $\bar{\mathcal{A}} \subseteq \mathcal{A}$; i.e., the tree uses a subset of the arcs in $\mathcal{A}$ and spans the nodes in $\mathcal{N}$. A *minimal spanning tree* (MST) of a connected graph $(\mathcal{N}, \mathcal{A})$ is a spanning tree whose cost is less than or equal to the cost of any other spanning tree of $(\mathcal{N}, \mathcal{A})$.

Let $\mathcal{Z} \subset \mathcal{N}$. A *Steiner tree* over $(\mathcal{N}, \mathcal{A}, \mathcal{Z})$ is a tree that uses nodes in $\mathcal{N}$, arcs in $\mathcal{A}$, and spans the nodes in $\mathcal{Z}$. Define $\mathcal{N} - \mathcal{Z}$ to be those nodes in $\mathcal{N}$ but not in $\mathcal{Z}$. Nodes in $\mathcal{N} - \mathcal{Z}$ used in the Steiner tree are called *Steiner points*. Figure 1.2 illustrates a spanning tree of an undirected graph, and also a Steiner tree with $\mathcal{Z} = \{A, B, C, D, E\}$. In the Steiner tree picture, the nodes $U$, $V$, and $W$ are Steiner points.

In telecommunications, the concept of an *interface* is important. An interface is the endpoint of a logical or physical communications channel between two nodes. For example, a 1 Gbps (gigabits per second) Ethernet port (physical point of connection on a router) might be divided into 5 interfaces, each

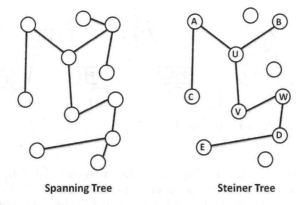

**Fig. 1.2** Spanning and Steiner trees

of bandwidth 200 Mbps (megabits per second), and these 5 interfaces might connect to different devices (e.g., interfaces 1-3 might connect to neighboring routers, while interfaces 4 and 5 might connect to subtending hosts). These different interfaces could be assigned different costs. A given arc $a$ connecting nodes $n_1$ and $n_2$ will have an interface (say $i_1$) on $n_1$ and an interface (say $i_2$) on $n_2$. Thus each interface on a node is an endpoint of a distinct arc originating or terminating at the node. The interface identifier is local to a node, so that two different nodes might both have an interface labelled $i_1$.

A *tunnel* in a network is a logical path between two nodes. Suppose $u \in \mathcal{N}$, $v \in \mathcal{N}$, but $(u, v) \notin \mathcal{A}$. Suppose we want specially marked packets (e.g., all packets with a specific source host address) that arrive at $u$ to take a specified tunnel to $v$. The path taken by this tunnel might be the shortest path between $u$ and $v$, or it might not be, e.g., if the tunnel is designed to utilize only special high bandwidth arcs. Each node on this tunnel, except for $v$, stores the knowledge of the next node on the tunnel. Multiple tunnels can be configured between the same pair of nodes, e.g., to provide enhanced security by sending traffic for different customers over different tunnels.

With a *Generic Routing Encapsulation* (GRE) tunnel [39], an *inner* packet is *encapsulated* in an *outer* packet at one end of the tunnel, and the outer packet is sent to the remote end. At the remote end, the encapsulation is removed and the inner packet is then processed or forwarded as required. With GRE tunnels, the inner packet is not processed by *via* routers (routers between the tunnel endpoints), and the tunnel is treated as any other arc by the IP routing protocol.

Another type of tunnel, called a *label switched path*, is illustrated in Figure 1.3. Suppose we want to tunnel between $u$ and $v$ using the bold arcs illustrated on the left. The tunnel can be created by encapsulating each packet that arrives at $u$ by placing it in another packet with an *outer label*. For example, in Figure 1.3, suppose that each packet arriving at $u$ with source host address $s$ is encapsulated and given the outer label $L_1$. Node $u$ consults its

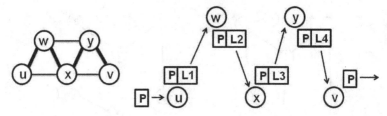

**Fig. 1.3** A label switched path tunnel

*label forwarding table* to learn that any packet with outer label $L_1$ should be sent to $w$. Upon arriving at $w$, the label forwarding table at $w$ specifies that, if a packet arrives at $w$ with outer label $L_1$, then $L_1$ should be swapped with (i.e., replaced by) $L_2$ and then the encapsulated packet should be sent to $x$. Similarly, at $x$, label $L_2$ is swapped for $L_3$, and the packet is sent to $y$. At $y$, label $L_3$ is swapped for $L_4$, and the packet is sent to $v$. The forwarding table at $v$ instructs it to now *decapsulate* any encapsulated packet with outer label $L_4$, at which point the decapsulated packet is available for further processing, e.g., to be forwarded to a directly attached receiver host.

We are now ready to define the multicast routing problem. We assume that each source host $s$ subtends the single node $n(s) \in \mathcal{N}$, and that each receiver host $r$ subtends the single node $n(r) \in \mathcal{N}$. We say that $s$ *subtends* or *is behind* node $n(s)$, and similarly for $n(r)$. Under this assumption, to route to a receiver host $r$, it suffices to route to $n(r)$. Similarly, to route from a source host $s$, it suffices to route from $n(s)$. We say that node $n \in \mathcal{N}$ *is used* by multicast group $g$ if there is a receiver host behind $n$ that has joined $g$, in which case we say that $n$ is a *receiver node* for $g$. Similarly, we say that a node $n \in \mathcal{N}$ is used by $g$ if there is a source host behind $n$ for $g$, in which case we say that $n$ is a *source node* for $g$. Each node will in general be used by multiple groups, and a given node can be both a source and receiver node for a group.

For group $g$, let $\mathcal{R}(g)$ be the set of receiver nodes for $g$, and let $\mathcal{S}(g)$ be the set of source nodes for $g$. For example, if $g$ is a broadcast group, then $\mathcal{R}(g) = \mathcal{N}$. Note that the definition of $\mathcal{R}(g)$ is independent of the identity of the source host sending to $g$; we will later consider SSM, for which the set of receiver nodes depends on $(s, g)$, that is, on both $g$ and the source host $s$ of packets for $g$. Let $\mathcal{G}$ be a set of multicast groups. The *multicast routing problem* is to determine, for each $g \in \mathcal{G}$ and each $s \in \mathcal{S}(g)$, a path from $s$ to all the nodes in $\mathcal{R}(g)$.

An alternative definition of this static problem was given by Waxman [110], who refers to a set of nodes to be interconnected in a *multipoint connection*, rather than to a source sending to a set of receivers. For a given group $g$ and source node $s$ used by $g$, define $\mathcal{Z} = \{s\} \cup \mathcal{R}(g)$. Then the static problem of connecting the nodes in $\mathcal{Z}$ with minimal cost is the problem of computing a minimal cost Steiner tree over $(\mathcal{N}, \mathcal{A}, \mathcal{Z})$. Note that our definition of the multicast routing problem assumes the set of arcs in the network is specified; we are not computing a network design/topology for a given set of nodes, e.g., as in [96]. Also, even though some applications require more bandwidth than others (e.g., video streaming compared with stock price information), we will not, in this primer, be concerned with the bandwidth of the arcs in the network and the bandwidth required by the multicast application.

Waxman [110] also states a dynamic version of this problem. The problem input data is a network, and a sequence of requests, where each request is to either add or delete a node from a multipoint connection. The objective is to determine a corresponding sequence of minimum cost trees, such that tree $k$ interconnects all the nodes in the multipoint connection following request $k$. If we allow the routing to be re-optimized after each request, this dynamic version is equivalent to a sequence of independent static problems.

The *unicast routing table* for node $n \in \mathcal{N}$ specifies, for each destination (a node or subnet or address prefix) known to $n$, the outgoing arc or interface on the shortest path from $n$ to that destination. Given two nodes $x$ and $y$ on a tree rooted at $s$, we say that $y$ is *downstream* of $x$ if $x$ is on the shortest path in the tree from $s$ to $y$. The *multicast routing table* specifies, for each group $g$ known to $n$, the set of outgoing arcs over which a replica of an incoming packet should be sent in order to reach the downstream receivers for $g$. For example, in Figure 1.1, suppose nodes $\{3, 4, 5, 8, 9, 10\}$ are receiver nodes for some $g$, and node 1 is the source node for $g$. Then the multicast routing table at node 2 for $g$ specifies the outgoing arcs $\{(2, 3), (2, 7)\}$ for packets arriving from node 1, the multicast routing table at node 3 for $g$ specifies the outgoing arcs $\{(3, 4), (3, 5)$ for packets arriving from node 2, and the table at node 7 for $g$ specifies the outgoing arcs $\{(7, 8), (7, 9), (7, 10)\}$ for packets arriving from node 2.

Although this primer is mostly concerned with networks with fixed position nodes and arcs, there is a large literature on *mobile ad-hoc networks* (MANETs). A MANET is set of mobile nodes which communicate over shared wireless channels, without any fixed network infrastructure or central network management. Such networks are used in, e.g., emergency search and rescue, and in military battlefields. Each mobile node participates in an ad hoc routing protocol by which it discovers nearby nodes and creates multihop paths through the network. Since the transmission range of each node is limited, multiple hops might be necessary to send a stream; each mobile node acts as both a host and a router that forwards packets. Since battery life is a major concern, multicast routing protocols for MANETs are often concerned with minimizing power consumption. A large number of methods have been proposed for multicast routing in a MANET; see Chen and Wu [21] and Junhai, Danxia, Liu, and Mingyu [56], which surveys 32 schemes and provides a taxonomy for grouping these methods. We review one MANET scheme in Section 4.6.

## 1.5 Broadcasting by Flooding

An easy but inefficient method for broadcasting is *flooding*. A simple flooding method starts with the source node sending a packet to each of its neighbors. Each node receiving the packet in turn forwards the stream to each of its neighbors, except that the packet is not forwarded to the neighbor from whom the packet was received. With such a scheme, it is necessary to prevent packets from cycling around the network, being continuously retransmitted. One way to accomplish this is to include a hop count in each packet. The source initializes this value to a large value, e.g., the network diameter as measured in hop count, with each arc cost set to 1. Each node processing the packet decrements the value by 1; when the value reaches 0, the packet is discarded. Since the diameter may not be known a priori, the upper bound $|\mathcal{N}| - 1$ can be used, where $|\mathcal{N}|$ is the number of elements in $\mathcal{N}$.

The above procedure can be improved by preventing a node from redistributing the multiple copies it may receive of a packet. For example, considering Figure 1.4, suppose the source $a$ must send a stream to nodes $b$, $c$, $d$, and $e$. Node $a$ will send a copy of each packet to $b$ and $c$, which in turn

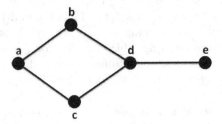

**Fig. 1.4** Improved flooding

send it to $d$, so $d$ receives two copies of each packet. However, $d$ needs to send only one copy of each packet to $e$. To accomplish this, the source can mark each copy of a given packet with the same packet identifier; each downstream node will then process only the first packet received with a given identifier, and discard the other packets with the same identifier. Having each node retain knowledge of the identifier of processed packets simultaneously prevents packets from cycling indefinitely, and prevents multiple copies of the same packet from being retransmitted by a given node [81]. (This rule is used in Content Addressable Network method described in Section 4.3.) Formally, this method is described as follows. Let $\mathcal{N}(n)$ be the set of nodes adjacent to $n$, i.e., reachable in one hop from $n$. First, the source node $s$ sends a packet to each of its neighbors. Each node $n$ receiving the packet executes the procedure in Figure 1.5. If there are multiple sources that wish to flood packets, the identifiers imposed by each source node must be globally unique. This can

---

**procedure** *Flood&Mark*()
1      **if** the packet was not previously processed by node $n$ {
2            Let $\bar{n}$ be the node from which the packet was received;
3            Send the packet to each $n \in \mathcal{N}(n)$ except $\bar{n}$;
4            Mark the packet as processed by $n$;
5      }
6      **else** discard the packet;

---

**Fig. 1.5** Flood&Mark

be achieved, e.g., by having the identifier imposed by $s$ contain the globally unique address of $s$.

Flooding explores all possible paths between the source and destination nodes, so a packet will reach a destination node as long as one path is available. This makes flooding robust and well-suited to networks prone to node or arc failures. However, flooding is bandwidth intensive, since each packet is retransmitted to all neighbors except for the neighbor from whom the packet was received [101].

## 1.5.1 Reverse Path Forwarding

The next multicast routing improvement we consider is *reverse path forwarding* (RPF), first proposed in 1977 by Dalal [27]. This technique is applicable to general multicast routing, not just broadcasting. Consider a source $s$ for group $g$ that is sending packets to a set of receivers that have joined $g$. Suppose some node $n$ receives a packet on arc $a$ and must decide what to do with the packet. The *reverse path forwarding* (RPF) check is: if the shortest path from $n$ to $s$ uses arc $a$, the check *passes*, and $n$ should forward the packet on each of the arcs specified in its multicast forwarding table for group $g$; otherwise, the check *fails* and $n$ should discard the packet. The unicast routing table at $n$ is used to determine if arc $a$ is on the shortest path from $n$ to $s$. If the unicast routing method has found more than one minimal cost route, a tie-breaking mechanism (e.g., a hash function) will select one of them.

What distinguishes RPF from unicast forwarding is that RPF forwarding depends on the *source* node, while unicast forwarding depends on the *destination* node. For example, consider Figure 1.6. Suppose the shortest path (using unicast routing) from $n$ to $s$ uses arc $a$. If $n$ receives a packet on the incoming arc $a$ then the RPF check passes, since $a$ is on the shortest path from $n$ to $s$; the packet is replicated at $n$ and sent out, say, on arcs $c$, $d$, and

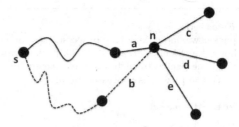

**Fig. 1.6** Basic RPF check

$e$, as specified in the multicast routing table at $n$ for $g$. The packet is not sent
back out over $a$, since a packet is never sent back out on the arc over which
it arrived. If $n$ receives a packet on the incoming arc $b$ then the RPF check
fails, since $b$ is not on the shortest path from $n$ to $s$; the packet is discarded.

As observed by Perlman [84], the RPF check, though quite simple, greatly
reduces the overhead of flooding. Since a node $n$ accepts a given packet only
on one incoming arc, then $n$ forwards this packet (on each of the interfaces
in its multicast forwarding table) only once, and the packet traverses any arc
in the network at most once.

With RPF, the decision whether to drop a packet is made only *after* the
packet has been replicated and sent to a neighboring node. Consider an arc $a$
between node $i$ and an adjacent node $j$. Suppose $a$ is not on the shortest path
from $j$ to the source. The RPF check says that $j$ should drop a packet arriving
from $i$ on $a$. Since the packet will be dropped by $j$, there is no reason for $i$
to send the packet to $j$. The *extended RPF check* of Dalal and Metcalfe [28]
says that $i$ should send a packet to a neighbor $j$ along arc $a$ only if $a$ is on the
shortest path from $j$ to the source. Node $i$ can easily determine if this holds in
a link state multicast routing protocol, or even in a distance vector protocol,
if each neighbor advertises its upstream next hop [98]. The extended RPF
check is illustrated in Figure 1.7, where the cost of four undirected arcs is
shown. The basic RPF check says that if $x$ receives from $w$ a packet originated

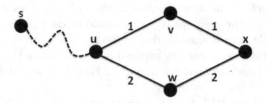

**Fig. 1.7** Extended RPF check

by $s$, the packet should be discarded since $(x, w)$ is not on the shortest path
to $s$. The extended RPF check says that $w$ should never forward the packet

to node $x$, since $(x, w)$ is not on the shortest path from $x$ to $s$. This extended RPF check, however, has not proved to be popular in practice.

## 1.5.2 RPF with Asymmetric Arc Costs

If the arc costs are asymmetric, which is often the case in the Internet, the shortest path from a source node $s$ to a receiver node $t$ will in general not be the same as the shortest path from $t$ to $s$. The validity of the RPF method for *broadcasting* does *not* require the assumption of symmetric arc costs [28]. However, the RPF method can cause *multicasting* to fail, unless algorithmic enhancements are made. This is illustrated in Figure 1.8 for the trivial case of one source node $s$ and one receiver node $t$. The arc $a_1$ has cost 1 in the direction $t \to s$ and cost 3 in the reverse direction, and the arc $a_2$ has cost 2 in both directions. Anticipating the PIM-based protocols discussed in

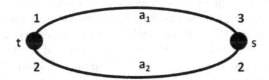

**Fig. 1.8** RPF with asymmetric costs

Section 3.7, suppose node $t$ sends, via shortest path unicast, a *join* request to $s$ that any packets sourced by a host subtending $s$ should be sent to $t$. This request will take arc $a_1$. When $s$ heeds the request and sends to $t$ using shortest path routing, packets will take arc $a_2$. If $t$ uses the RPF check, it will discard all packets sent by $s$. There are at least two solutions to this dilemma.

One solution is for $t$ to know the incoming interface that will be taken by packets following the shortest path from $s$ to $t$. This approach, proposed in 2008 by Li, Mirkovic, Ehrenkranz, Wang, Reiher, and Zhang [66], requires the join request sent from a receiver node $t$ to a source node $s$ to follow the same path that $s$ would use to send to $t$ using shortest path routing. Implementation of this method requires each node $n$ to store, for each entry in its forwarding table, the *forward* interface into $n$ that would be used by packets taking the shortest path from that entry (now treated as a source) to $n$. The request sent by $n$ to join a tree rooted at some source node $s$ can now be sent along a sequence of forward interfaces, and this path (the true shortest path from $s$ to $n$) is used by $s$ to send packets down to $n$. (This method can also be used to detect *spoofing*, where the source of a packet is deliberately forged, e.g., to create denial of service attacks.)

A second approach takes the opposite tack, and requires that the path taken by packets sent from a source node $s$ to a receiver node $t$ take the same path as the shortest path from $t$ to $s$. This is the approach adopted by PIM-based multicast protocols. The drawback of this approach is that, if the arc costs are asymmetric, the path taken by packets sent from $s$ to $t$ may not be the shortest path from $s$ to $t$; the path is only guaranteed to be the shortest path from $t$ to $s$. This possibility of non-optimal routing has not prevented PIM-based protocols from being widely adopted.

## 1.6 The History of Multicast

As the reader now has a glimpse into multicast, we provide a very few words on its history. The groundwork for Internet multicast was laid by Deering in his 1991 Ph.D. dissertation [32]. The first large scale multicast deployment began in 1992, with the MBone (Multicast Backbone). The MBone used unicast encapsulated tunnels to connect a set of small multicast networks. These tunnels allowed multicast traffic to traverse Internet routers that were not multicast enabled. IP Multicast traffic was encapsulated upon entry to a tunnel, and sent using unicast to the destination end of the tunnel, where it was decapsulated. The MBone grew from 40 subnets in 4 countries in 1992 to 2800 subnets in 25 countries by 1996 [98]. In 1997, work began on a hierarchical approach to inter-domain multicast. In this approach, different multicast protocols could be used in each domain, domains exchange information on multicast sources, and intra-domain trees are connected by an inter-domain tree. By 1999, there were two Internet2 backbone multicast networks, vBNS and Abilene, which began to supplant the MBone.

In the 1990s there was intense focus on multicast routing methods (see the 1997 review by Diot, Dabbous, and Crowcroft [35], and the 2000 review by Almeroth [6]). In a 2006 survey by Oliveira, Pardalos, and Resende [81] of optimization problems related to multicast, 64% of the references were to papers published between 1992 and 1998, and only 18% are 2000 or later. However, multicast has enjoyed a resurgence of interest, due the ability of service providers to now offer reliable, scalable multicast in the Internet and in Virtual Private Networks (Chapter 7).

# Chapter 2
# Basic Concepts in Tree Based Methods

Tree based approaches distribute packets along a tree. The tree is logical, but not necessarily physical; the routing follows a tree but the underlying physical network topology (e.g., the fiber optic cables) need not have a tree topology. Each node maintains, for each multicast group $g$, a multicast routing table for $g$, which specifies which interfaces are part of the tree for $g$. For example, consider Figure 2.1, which shows the tree for a particular $g$. The solid arcs are in the tree, and the dotted arcs are in the physical net-

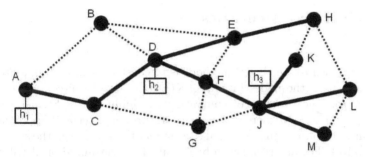

**Fig. 2.1** Multicast tree

work, but not in the tree. Suppose $h_1$, $h_2$, and $h_3$ are both receiver hosts and source hosts for $g$. Let $\mathcal{A}_n(g)$ denote the interface list at node $n$ for group $g$. We have $\mathcal{A}_A(g) = \{(A,C),(A,h_1)\}$, $\mathcal{A}_B(g) = \emptyset$, $\mathcal{A}_C(g) = \{(C,A),(C,D)\}$, $\mathcal{A}_D(g) = \{(D,C),(D,E),(D,F),(D,h_2)\}$, etc. If $A$ receives a packet from $h_1$ and the RPF check passes, then $A$ examines $\mathcal{A}_A(g)$ and sends the packet out $(A,C)$. If $A$ receives a packet from $C$ and the RPF check passes, then $A$ sends it out on $(A,h_1)$. Similarly, if $D$ receives a packet from $h_2$ and the RPF check passes, then $D$ sends it out on $(D,C)$, $(D,E)$, and $(D,F)$, while if $D$ receives a packet from $E$ and the RPF check passes, $D$ sends it out on $(D,C)$, $(D,F)$, and $(D,h_2)$. Thus a packet for $g$ arriving at node $n$ which passes the RPF check is replicated and sent out on each interface in $\mathcal{A}_n(g)$ except the incoming interface.

Due to practical considerations, use of a tree does not eliminate the need for an RPF check. Consider Figure 2.2, where $S$ is the source node. Suppose convergence of the unicast routing tables has not yet occurred, so that $A$ thinks that $B$ is its *RPF neighbor* (i.e., the next node on the shortest path from $A$ to the source node $S$), $B$ thinks that $C$ is the RPF neighbor, and $C$

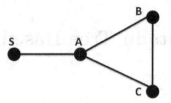

**Fig. 2.2** RPF check with nonconvergence

thinks that $A$ is the RPF neighbor. Then any packets sourced by $A$ caught in the $A \rightarrow C \rightarrow B \rightarrow A$ loop will cycle until they time out. However, until convergence has occurred, the RPF check at $A$ will prevent any new packets from $S$ from entering the $A \rightarrow C \rightarrow B \rightarrow A$ loop [90].

## 2.1 Steiner Tree Heuristics

One approach to tree based methods is to construct, for each $g$, a Steiner tree which spans the nodes in $\mathcal{S}(g) \cup \mathcal{R}(g)$ (the set of source and receiver nodes for $g$). The problem of computing a minimum cost Steiner tree is NP-complete [42], and there is a large literature on exact and approximate algorithms for constructing minimal cost Steiner trees (see, e.g., the survey by Winter [113]). The algorithms can be classified as centralized or distributed. Centralized methods incorporate full information about the topology of the network. In decentralized methods each node independently makes routing decisions based upon a limited or summarized view of the network topology. Centralized methods typically utilize either Prim's or Kruskal's MST method. In Prim's 1957 method, the tree is initialized as any given node. In each iteration the node closest to the tree, and the arc from the tree to that node, are added to the tree; the method terminates when all nodes have been added to the tree. In Kruskal's 1956 method, each node is initially its own subtree, and in each iteration the lowest cost arc connecting two subtrees is added, until all nodes are in a single tree [5].

A popular centralized heuristic for building Steiner trees is the method of Kou, Markowsky, and Berman (KMB) [64]. Given the undirected graph $\Psi = (\mathcal{N}, \mathcal{A})$ and a set $\mathcal{Z} \subset \mathcal{N}$ of nodes to interconnect, Step 1 is to consider the auxiliary complete undirected graph $\Delta = (\mathcal{Z}, \mathcal{E})$, where there is one arc in $\mathcal{E}$ for each (unordered) pair of nodes in $\mathcal{Z}$. The cost of $(i, j) \in \mathcal{E}$ is the cost of the shortest path, in the original graph $\Psi$, between nodes $i$ and $j$. Corresponding to each arc $(i, j) \in \mathcal{E}$ is a path in $\Psi$ between nodes $i$ and $j$. Step 2 is to compute an MST of $\Delta$ (if the MST is not unique, pick one arbitrarily). Step 3 is to construct a subgraph of $\Psi$ by replacing each edge in the MST by its corresponding shortest path in $\Psi$. Step 4, the final step, is to create a Steiner tree $T$ by deleting arcs from the subgraph so that all leaf nodes of $T$ belong to $Z$. It can be shown [64] that the cost of the Steiner tree computed by this method is no more than twice the cost of the minimal cost Steiner tree.

The method is illustrated in Figure 2.3. Step (0) shows the starting graph $\Psi$, and $Z = \{A, E, F\}$. Next to each arc is the arc cost. Step (1) shows the auxiliary graph $\Delta$ with the shortest path distance between each pair of nodes in $Z$. In Step (2), the MST of $\Delta$ is chosen. In Step (3), the arc $(A, E)$ in $\Delta$ corresponds to the path $A \rightarrow B \rightarrow D \rightarrow E$ in $\Psi$ (other paths with cost 4 could have been chosen), and arc $(A, F)$ in $\Delta$ corresponds to the path $A \rightarrow C \rightarrow D \rightarrow F$ in $\Psi$ (other paths could have been chosen). An MST of the subgraph is chosen in Step (4), and the final pruning is done in Step (5). For this example, the method yields an optimal (minimal cost) Steiner tree.

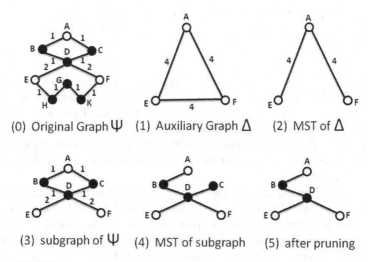

(0) Original Graph $\Psi$    (1) Auxiliary Graph $\Delta$    (2) MST of $\Delta$

(3) subgraph of $\Psi$    (4) MST of subgraph    (5) after pruning

**Fig. 2.3** KMB Steiner tree

Waxman [110] provides an example where, for this method, the worst case bound of twice the optimal cost is approached. In Figure 2.4, top picture, the nodes $1, 2, \cdots, k$ are to be interconnected. For $0 < \theta << 1$, the optimal tree is shown in the middle picture, while the lower picture shows the tree generated by the KMB method. Since $[2x(k-1)]/[k(x+\theta)] \rightarrow 2$ as $k \rightarrow \infty$ and $\theta \rightarrow 0$, the worst case ratio of 2 is asymptotically achieved. Waxman also compares the KMB method against other Steiner tree heuristics, using a random graph model in which an arc between two nodes is created with probability $\beta e^{-d/(L\alpha)}$, where $d$ is the distance between the two nodes, $L$ is the maximum distance between any two nodes, $\alpha \in (0, 1]$, and $\beta \in (0, 1]$. Increasing $\beta$ yields more arcs, while decreasing $\alpha$ increases the frequency of short arcs relative to longer arcs. (This random model has been utilized extensively by subsequent researchers.)

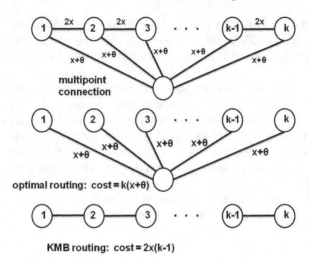

**Fig. 2.4** Worst case for KMB heuristic

While the KMB method is quite useful in some applications where the set of nodes used by the group is static (e.g., computer aided design of integrated circuits and printed wire boards [94]), the method is not particularly useful in telecommunications, where sources and receivers come and go. However, as observed by Waxman, methods such as KMB provide benchmarks against which fast heuristics, suitable for dynamic telecommunications networks, can be compared.

There are many variants of the Steiner tree problem. Haḉ and Zhou [45] consider minimum cost Steiner trees subject to bounds on the delay from the source to each destination. Another method for computing delay constrained trees is discussed in Section 4.5 below. Secci et al. [97] study a variant which can be used by service providers to model inter-Autonomous System routing policies. Chaintreau et al. [19] consider the impact of queuing delay in multicast trees, and show that the throughput decreases significantly as a function of the number of receivers. The multicast packing problem [81] is to determine a tree for each group such that, given the bandwidth required by each group, the maximum bandwidth, over all arcs in the network, is minimal. Steiner tree methods can also be classified as either solving a static problem, for which the set of sources and receivers is known a priori, or solving a dynamic problem, for which the set of sources and receivers is not known a priori, and receivers can join or leave the group at any time. These methods are surveyed by Novak, Rugelj, and Kandus [80].

As applied to distributing packets for a multicast group $g$, the KMB method does not distinguish between source and receiver nodes, and simply constructs a tree which spans any node used by $g$. Such a tree is called a *shared tree*. We will return to the subject of shared trees, but first we consider another type of tree, called a *source* tree.

## 2.2 Source Trees

Consider a group $g$, and a source host $s$ for $g$, where $s$ subtends the source node $n(s)$. A *source tree* for $(s, g)$ has $n(s)$ as the root of the tree, and spans the set of nodes wishing to receive the $(s, g)$ stream, i.e., the stream sent by $s$ to group $g$. A source tree is in general also a Steiner tree, since it may contain nodes not used by $g$. Packets only flow down the tree (away from the root), from the source to the receivers, and never up a source tree towards the root. The source tree is identified by the pair $(s, g)$.

For a rooted tree, the *upstream neighbor*, also called the *RPF neighbor*, of node $n$ is the neighbor of $n$ that is on the shortest path from $n$ to the root. For a source tree, the root is the source node. Referring to Figure 2.5, suppose for some group $g_1$ there are source hosts $s_1$ and $s_2$ behind node 1, and suppose that behind nodes 3, 4, 6, 7, and 9 are hosts wishing to receive the $(s_1, g_1)$ stream. The $(s_1, g_1)$ source tree is rooted at node 1, and is drawn using heavy solid lines. The upstream neighbor of node 7 is node 3, the upstream neighbor of node 3 is node 2, and the upstream neighbor of node 2 is the source node 1. At node 1 the *outgoing interface list* (OIL) for

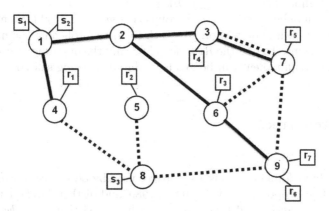

**Fig. 2.5** Source tree

$(s_1, g_1)$ is $\mathcal{A}_1(s_1, g_1) = \{(1, 2), (1, 4)\}$, so packets sourced by $s_1$ are replicated at node 1, and send to nodes 2 and 4. At node 2 we have $\mathcal{A}_2(s_1, g_1) = \{(2, 3), (2, 6)\}$. At node 6 one we have $\mathcal{A}_6(s_1, g_1) = \{(6, 9), (6, r_3)\}$. At node 9, we have $\mathcal{A}_9(s_1, g_1) = \{(9, r_6), (9, r_7)\}$. At each node on the source tree, the RPF check is used to determine whether an incoming packet should be replicated and sent over all arcs in the OIL, or the packet should be discarded.

Each node on the $(s_1, g_1)$ tree is said to maintain *state* for $(s_1, g_1)$. In Figure 2.5, nodes 1, 2, 3, 4, 6, 7, and 9 each maintain $(s_1, g_1)$ state. Note that node 2 maintains $(s_1, g_1)$ state even though there are no subtending receiver hosts wanting the $(s_1, g_1)$ stream. The state information includes the interface leading to the RPF neighbor, and the set of downstream arcs, as specified by the OIL. For example, node 2 stores the upstream arc $(2, 1)$ leading to the RPF neighbor (node 1), and $\mathcal{A}_2(s_1, g_1)$.

Now suppose that, for the same group $g_1$, there is another source host $s_2$ behind node 1, and that behind nodes 3, 4, 6, 7, and 9 are hosts wishing to receive the $(s_2, g_1)$ stream. We now create another source tree, the $(s_2, g_1)$ tree. This source tree might use exactly the same set of arcs as the $(s_1, g_1)$ tree, or it could be designed using different methods, and not use the same set of arcs (e.g., if the tree construction method is bandwidth aware, and the streams generated by $s_1$ and $s_2$ have different bandwidths). This new source tree is an $(s_2, g_1)$ tree, $(s_2, g_1)$ state is created on each node on this tree, and each node stores the interface to the RPF neighbor and the $(s_2, g_1)$ OIL.

Suppose now that host $s_3$, subtending node 8, is the source for a different group $g_2$, and the receiver nodes for $(s_3, g_2)$ are 3, 4, 5, 6, and 7. The arcs for a $(s_3, g_2)$ source tree, rooted at node 8, are shown in heavy dotted lines. We have $\mathcal{A}_8(s_3, g_2) = \{(8, 4), (8, 5), (8, 9)\}$, $\mathcal{A}_9(s_3, g_2) = \{(9, 7), (9, r_6), (9, r_7)\}$, and similarly for the other nodes on the $(s_3, g_2)$ tree. The RPF neighbor of node 8 is itself (it is the root), the RPF neighbor of node 9 is node 8, etc. Each node on the $(s_3, g_2)$ tree maintains $(s_3, g_2)$ state, and stores the interface to the RPF neighbor and the $(s_3, g_2)$ OIL.

When the number of sources and groups is large, considerable memory can be required at each node to store, for each $(s, g)$ tree touching the node, the $(s, g)$ pair, the interface to the RPF neighbor, and the outgoing interface list. The memory requirement can be greatly reduced by using shared trees.

## 2.3 Shared Trees

A *shared tree* for a group $g$ is a single tree used by *all* sources for $g$. Shared trees were first proposed by Wall [109] in 1980, and further developed in 1993 by Ballardie, Francis, and Crowcroft [10] who called them *core based trees* (CBTs). The incarnation of shared trees popular today is *Prototcol Independent Multicast - Sparse Mode* (PIM-SM), developed by Estrin et al. in 1997 ([37], [38]). CBTs are studied in Section 3.6 and PIM-SM in Section 3.7.2. Another type of shared tree, BiDirectional PIM, is studied in Section 3.7.3. Shared trees mitigate some of the scaling problems (consumption of memory, bandwidth, or node processing resources) of source trees.

Since there is no source node to serve as the root of a shared tree, one privileged node serves at the root. For CBTs, this node is called a *core*; for PIM-SM it is called a *rendezvous point* (RP). An RP and a core serve the same function. They are typically chosen in the middle of the geographic region spanned by the group sources and receivers (e.g., if the multicast group spans the U.S., the core or RP might be in Kansas). Figure 2.6 illustrates, using heavy lines, a shared tree, rooted at node 5, for a group $g$. There are

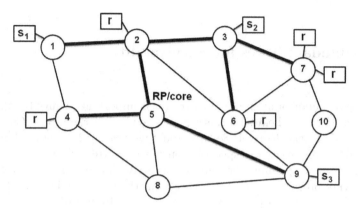

**Fig. 2.6** Shared tree

source hosts for $g$ behind nodes 1, 3, and 9, and receiver hosts behind nodes 2, 4, 6, and 7. Since a shared tree for group $g$ is used for any source host for $g$, the tree is represented by the pair $(\star, g)$, where $\star$ denotes the "wild card," meaning it matches any source host for the group. Thus $(\star, g)$ state appears on each node spanned by the shared tree, whether or not a source host or receiver host subtends the node. We have $(\star, g)$ state on all nodes except 8 and 10. The RPF neighbor of node 7 is node 3, the RPF neighbor of node 3 is node 2, and the RPF neighbor of node 2 is node 5. Similarly, the RPF neighbor of node 9 is node 5, etc.

One disadvantage of shared trees, compared to source trees, is the potential for non-optimal routing. The example in Figure 2.4 above illustrates this, since for the tree generated by the KMB method (lower figure), the worst case cost between any two nodes is $2x(k - 1)$, while for small $\theta$ the worst case cost in the optimal Steiner tree (middle figure) is $2(x + \theta)$. Even with a minimal cost shared tree, non-optimal routing can occur, as the following simple example demonstrates. Consider a group $g$ and three nodes $a$, $b$, and $c$, at the vertices of an equilateral triangle, with unit arc costs. Suppose there is a source host and a receiver host for $g$ behind each node. Suppose the $(\star, g)$ shared tree is $\{(a, b), (b, c)\}$. Then each node has the single state $(\star, g)$. Traffic from $a$ to $c$ will follow the path $a \rightarrow b \rightarrow c$, with cost 2. If instead we use source trees, we would generate three trees, one rooted at each node, and each node would have three $(s, g)$ states. Packets from $a$ to $c$ would follow the path $a \rightarrow c$ on the tree rooted at $a$, with a path cost of 1.

As a final example, referring to Figure 2.6 above, suppose all arc costs are 1. Then the optimal route from $s_3$ to the receiver host behind node 6 is the single hop $(9, 6)$. However, the route on the shared tree is the path $9 \rightarrow 5$ to reach the RP/core, and then the path $5 \rightarrow 2 \rightarrow 3 \rightarrow 6$ down the shared tree to reach node 6. This example also illustrates that the location of the core may greatly impact the path length and latency (i.e., delay) for a multicast stream.

## 2.4 Redundant Trees for Survivability

In case of a node or arc failure in a tree, it is important to quickly establish a new tree. Various schemes for this are surveyed by Bejerano, Busi, Ciavaglia, Hernandez-Valencia, Koppol, Sestito, and Vigoureux [12]. One method, applicable to source trees, is to specially construct two trees, rooted at the same source node $s$ and spanning the same set of receiver nodes. The two trees have the following property: if a node (other than the source) fails, then for any other node $n$ there is a path, on one of the trees, from $s$ to $n$. The method described in [12] has two main steps. First, a subgraph $\Psi$ is created which contains two node disjoint paths from the root to every other node. Second, two directed trees, with the desired property, are constructed from arcs in $\Psi$.

The desired subgraph $\Psi$ can be constructed from the original undirected graph $(\mathcal{N}, \mathcal{A})$ as follows. The initial step is to construct an *ear*, which is a cycle of length at least 3 containing the source $s$. So create any such cycle, and traversing the cycle in an arbitrary direction from $s$, let $\mathcal{S}^1$ be the ordered set of nodes in the cycle. For example, in Figure 2.7, sub-figure $(i)$, where $s = 0$, we let $\mathcal{S}^1 = \{0, 1, 4, 0\}$. If $\mathcal{S}^1 = \mathcal{N}$ we are done, since the cycle contains two disjoint paths from $s$ to each node on the cycle. Otherwise, pick a node

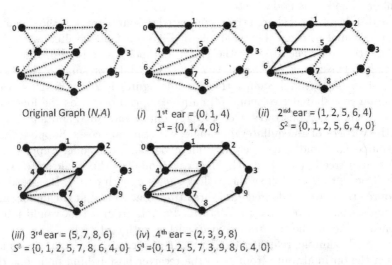

Original Graph $(N,A)$      $(i)$  1st ear = $(0, 1, 4)$      $(ii)$  2nd ear = $(1, 2, 5, 6, 4)$
                                   $S^1 = \{0, 1, 4, 0\}$             $S^2 = \{0, 1, 2, 5, 6, 4, 0\}$

$(iii)$  3rd ear = $(5, 7, 8, 6)$      $(iv)$  4th ear = $(2, 3, 9, 8)$
$S^3 = \{0, 1, 2, 5, 7, 8, 6, 4, 0\}$      $S^4 = \{0, 1, 2, 5, 7, 3, 9, 8, 6, 4, 0\}$

**Fig. 2.7** Redundant trees - node ordering

$n \in \mathcal{N} - \mathcal{S}^1$ and construct two node disjoint paths from $s$ to $n$, using, e.g., the classic method of Surballe [103]. Let $u$ be the node in $\mathcal{S}^1$ that is on one of the disjoint paths from $s$ to $n$, such that $u$ is closest to $n$. Let $v$ be the node in $\mathcal{S}^1$ that is on the other disjoint path from $s$ to $n$, such that $v$ is closest to $n$. Without loss of generality, assume $\mathcal{S}_u^1 \leq \mathcal{S}_v^1$, where $\mathcal{S}_u^1 = j$ if $u$ is in position $j$ of the ordered set $\mathcal{S}^1$. (It might be that $\mathcal{S}_u^1 = \mathcal{S}_v^1$ if, for example, $u = v = s$.) Starting from $u$, we traverse the path from $u$ to $n$ to $v$, and suppose, excluding $u$ and $v$, this path encounters the ordered set of nodes $\mathcal{Q}^2$. ($\mathcal{Q}^2$ is the *ear* connecting $n$ to $\mathcal{S}^1$.) We insert the ordered set $\mathcal{Q}^2$ into $\mathcal{S}_1$ just before position $v$, yielding the new ordered set $\mathcal{S}^2$. In Figure 2.7, sub-figure (*ii*), we select $n = 6$, yielding $u = 1$, $\mathcal{S}_u^1 = 2$ (i.e., "1" is in the second position of the array $\mathcal{S}^1$), $v = 4$, $\mathcal{S}_v^1 = 3$ (i.e., "4" is in the third position of the array $\mathcal{S}^1$), and $\mathcal{Q}_2 = \{2, 5, 6\}$; inserting $\mathcal{Q}^2$ before $v$ in $\mathcal{S}^1$ yields $\mathcal{S}^2 = \{0, 1, 2, 5, 6, 4, 0\}$. If $\mathcal{S}^2 = \mathcal{N}$ we are done, otherwise we select a node $n \in \mathcal{N} - \mathcal{S}^2$, create an ear $\mathcal{Q}_3$ connecting $n$ to $\mathcal{S}^2$, etc.

Continuing the example, in sub-figure (*iii*) we select $n = 8$ yielding $u = 5$, $\mathcal{S}_u^2 = 4$, $v = 6$, $\mathcal{S}_v^2 = 5$, $\mathcal{Q}^3 = \{7, 8\}$; inserting $\mathcal{Q}^2$ before $v$ in $\mathcal{S}^2$ yields $\mathcal{S}^3 = \{0, 1, 2, 5, 7, 8, 6, 4, 0\}$. Finally, in sub-figure (*iv*) we select $n = 9$ yielding $u = 2$, $\mathcal{S}_u^3 = 3$, $v = 8$, $\mathcal{S}_v^3 = 6$, $\mathcal{Q}^3 = \{3, 9\}$; inserting $\mathcal{Q}_3$ before $v$ yields $\mathcal{S}^4 = \{0, 1, 2, 5, 7, 3, 9, 8, 6, 4, 0\}$. Let $\mathcal{S}^\star$ be the final ordered set (containing all the nodes) created by this procedure. Then $|\mathcal{S}^\star| = N + 1$, where $N = |\mathcal{N}|$, since the root $s$ appears in the first and last positions, and all other nodes appear exactly once. We now use $\mathcal{S}^\star$ to create the two directed trees.

We create the first directed tree $\mathcal{T}_1$ as follows. For $n \in \mathcal{N}$, $n \neq s$, pick one node $p_n$ such that the undirected arc $(p_n, n)$ exists in the original graph $(\mathcal{N}, \mathcal{A})$, and such that $p_n$ appears *after* $n$ in the ordered set $\mathcal{S}^\star$; add the directed arc $(p_n, n)$ to $\mathcal{T}_1$. Continuing our example, Figure 2.8 shows how

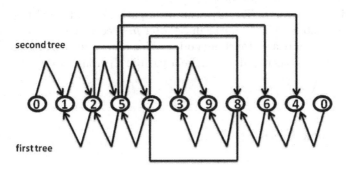

**Fig. 2.8** Redundant trees - tree creation

the indices $p_n$ are selected to create $\mathcal{T}_1$. In the center of the figure are the elements of the ordered set $\mathcal{S}^\star$, and below the node numbers we show, for each $n$, $n \neq s$, the selection of $p_n$ used to create $\mathcal{T}_1$. Starting from the next to last position of $\mathcal{S}^\star$, for node 4 we pick $p_4 = 0$, yielding the directed arc $(0, 4)$;

for node 6 we pick $p_6 = 4$, yielding the directed arc $(4, 6)$; for node 8 we pick $p_8 = 6$, yielding the directed arc $(6, 8)$; and we continue this way, generating the directed arcs $(8, 9)$, $(9, 3)$, $(8, 7)$, $(7, 5)$, $(5, 2)$, and $(2, 1)$. Note that the rule for creating this tree may allow multiple choices for $p_n$ for some $n$, so $T_1$ is not uniquely determined. Figure 2.9, sub-figure $(i)$ shows $T_1$.

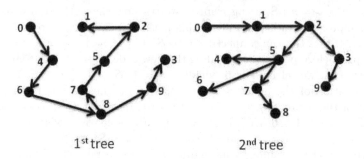

1st tree                                2nd tree

**Fig. 2.9** Redundant trees - final step

We create the second directed tree $T_2$ as follows. For each node $n \in \mathcal{N}$, $n \neq s$, pick one node $p_n$ such that the undirected arc $(p_n, n)$ exists in the original graph $(\mathcal{N}, \mathcal{A})$, and such that $p_n$ appears *before* $n$ in the ordered set $\mathcal{S}^\star$; add the directed arc $(p_n, n)$ to $T_2$. In Figure 2.8, above the node numbers we show, for each $n$, $n \neq s$, the selection of $p_n$ used to create $T_2$. Starting from the second position of $\mathcal{S}^\star$, for node 1 we pick $p_1 = 0$, yielding the directed arc $(0, 1)$; for node 2 we pick $p_2 = 1$, yielding the directed arc $(1, 2)$; and we continue this way, generating the directed arcs $(2, 5)$, $(5, 7)$, $(2, 3)$, $(3, 9)$, $(7, 8)$, $(5, 6)$, and $(5, 4)$. The tree $T_2$ also is in general not uniquely determined. Figure 2.9, sub-figure $(ii)$ shows $T_2$.

Consider any node $n$. Tree $T_1$ contains a path from $s$ to $n$ that uses only arcs touching nodes which appear in $\mathcal{S}^\star$ after $n$. Tree $T_2$ contains a path from $s$ to $n$ that uses only arcs touching nodes which appear in $\mathcal{S}^\star$ before $n$. Hence, if some node fails, there will either be a path in $T_1$ from $s$ to $n$, or a path in $T_2$ from $s$ to $n$.

## 2.5 Network Coding

In multicast routing, a node receives a packet on an incoming interface, and replicates the packet, sending it on a set of outgoing interfaces. In unicast routing, there is a single outgoing interface, so no replication is needed. With both approaches, the payload in each packet is unchanged, and only the routing information (e.g., the outgoing interface list) changes as the packet is forwarded. With *network coding*, introduced in a seminal 2000 paper by

Ahlswede, Cai, Li, and Yeung [4], a node receiving packets over several in-coming interfaces generates new packets by operating on the bits received over these incoming interfaces. Network coding can provide improved band-width utilization, lower delay, and, for wireless networks, reduced number of transmissions.

### 2.5.1 Examples of Network Coding

The following three examples from Sprintson [100] illustrate these benefits. In all these examples, we assume that all packets have the same fixed length, that all arcs have unit capacity (each arc can send one packet per unit time), that there is no delay for bits to traverse a node, and that nodes have infinite capacity.

Considering first Figure 2.10, sub-figure ($i$), suppose that source $s_1$ wants

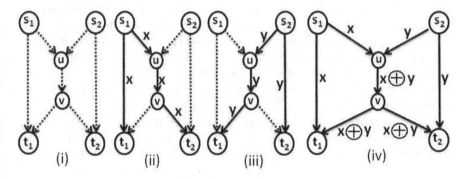

**Fig. 2.10** Network coding - first example

to send packet $x$ to $t_1$ and $t_2$, and source $s_2$ wants to send packet $y$ to $t_1$ and $t_2$. The only directed tree available to $s_1$ is illustrated in solid lines in sub-figure ($ii$), and the only directed tree available to $s_2$ is illustrated in solid lines in sub-figure ($iii$). However, $s_1$ and $s_2$ cannot both use these trees simultaneously, since arc $(u, v)$ has capacity 1. Using network coding, as shown in sub-figure ($iv$), node $u$ computes $x \oplus y$, which is defined as follows. Assume the packet is $k$ bits long. Then $x \oplus y$ is the packet of length $k$ whose $i - th$ bit is $x_i \oplus y_i$, where $x_i \oplus y_i = 1$ if $x_i + y_i = 1$ and 0 otherwise. Node $u$ sends $x \oplus y$ to $v$, who sends it to $t_1$ and $t_2$. Destination $t_1$ can recover $y$ from $x$ and $x \oplus y$ since, e.g., if $x_i = 1$ and $x_i \oplus y_i = 0$, then $y_i = 1$. Similarly, $t_2$ can recover $x$ from $y$ and $x \oplus y$. So, with network coding, $s_1$ can send $x$ over the tree rooted at $s_1$ at the same time $s_2$ sends $y$ over the tree rooted at $s_2$, even though the trees share a common arc.

The second example, shown in Figure 2.11, illustrates delay reduction using network coding. We define the *depth* of a tree to be the number of arcs in the longest path from the root to a leaf. Considering sub-figure (*i*), suppose

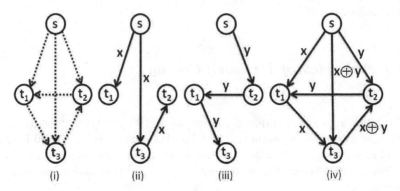

**Fig. 2.11** Network coding - second example

source $s$ must send packets $x$ and $y$ to destinations $t_1$, $t_2$, and $t_3$. Suppose $s$ uses the tree in sub-figure (*ii*) to send $x$ and the tree in sub-figure (*iii*) to send $y$. Since the depth of the first tree is 2 and the depth of the second tree is 3, it takes three time units for the three destinations to receive both $x$ and $y$ (this holds for any choice of arc disjoint trees). With network coding as in sub-figure (*iv*), only two time units are required for the three destinations to receive either $x$ and $y$, or $x$ and $x \oplus y$, or $y$ and $x \oplus y$.

The last example shows how network coding can reduce the number of wireless transmissions. Considering Figure 2.12, sub-figure (*i*), suppose $u$ must send packet $x$ to $v$, and $v$ must send packet $y$ to $u$. Without network

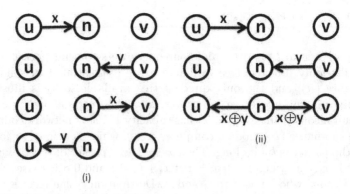

**Fig. 2.12** Network coding - third example

coding, four transmissions are required. With network coding, as shown in sub-figure $(ii)$, only three transmissions are required. Note that, since this is wireless transmission, it counts as one transmission when $n$ sends $x \oplus y$ to both $u$ and $v$ in sub-figure $(ii)$.

## 2.5.2 Theory of Linear Network Codes

The following discussion is based upon Sprintson [100]. The theory utilizes the rules of arithmetic over a finite field $GF(q)$. For our purposes, it suffices to let $GF(q)$ be the set of nonnegative integers less than $q$, for some prime number $q$, so if for some integers $\alpha$ and $\beta$ we have $0 \leq \alpha < q$ and $0 \leq \beta < q$, then $\alpha + \beta$ is defined to be their sum (mod $q$).

Consider a directed tree rooted at $s$, and suppose $s$ needs to send $k$ packets $(p_1, \cdots, p_k)$ to $D$ destination leaf nodes. Without loss of generality, we can assume that each of the $D$ destination nodes has $k$ incoming arcs (if this does not hold for some destination $d$, we can create a new destination $\bar{d}$ and $k$ directed arcs from $d$ to $\bar{d}$).

For a given node $n$, let $\mathcal{I}(n)$ be the set of incoming arcs at $n$ and let $\mathcal{J}(n)$ be the set of outgoing arcs at $n$. We reserve the first $k$ arc indices for the arcs incoming to $s$, so $\mathcal{I}(s) = \{1, 2, \cdots, k\}$. If for $n \in \mathcal{N}$ we have $|\mathcal{I}(n)| > 1$ (i.e., more than one incoming interface), then we select $|\mathcal{I}(n)| \cdot |\mathcal{J}(n)|$ coefficients $f_{ij}$ such that $i \in \mathcal{I}(n)$, $j \in \mathcal{J}(n)$, and $f_{ij} \in GF(q)$. For example, if $q = 2$ the field is $GF(2)$, and if $n$ has 3 incoming interfaces and 2 outgoing interfaces, then we generate 6 binary coefficients for this $n$. If $n$ has exactly one incoming arc $i$, each arriving packet is duplicated and sent out over each outgoing arc; in this case we set $f_{ij} = 1$ for $j \in \mathcal{J}(n)$. This coefficient generation is done for each node $n$. We denote by $\{f_{ij}\}$ the set of all the coefficients for all the nodes.

Suppose node $n$ receives packet $p_i$ on incoming arc $i \in \mathcal{I}(n)$, and let $p_j$ be the packet sent on outgoing arc $j \in \mathcal{J}(n)$. Then, for $j \in \mathcal{J}(n)$, the *linear network code* generated by the coefficients $\{f_{ij}\}$ is defined as

$$p_j = \sum_{i \in \mathcal{I}(n)} f_{ij} p_i \, ,$$

where the arithmetic operations are performed componentwise on each packet, and arithmetic is modulo $q$. Thus, for example, referring to Figure 2.13, where each arc is numbered, the packet $p_3$ sent out from $s$ on arc 3 is $f_{13}p_1 + f_{23}p_2$, the packet $p_4$ sent out from $s$ on arc 4 is $f_{14}p_1 + f_{24}p_2$, and the packet $p_9$ sent out from $c$ on arc 9 is $f_{69}p_6 + f_{79}p_7 = f_{69}p_3 + f_{79}p_4$. Since the packet sent on each outgoing arc from $n$ is a linear combination, using the $\{f_{ij}\}$ coefficients, of the packets on the incoming arcs to $n$, it follows that the packet sent on each arc leaving a node is a linear combination of the $k$ packets received by the source node $s$. In our example, we have, for the arcs 5 and 10 into $t_1$,

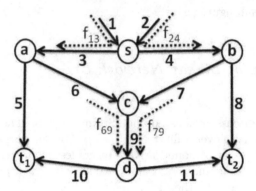

**Fig. 2.13** Network coding - coefficients

$$p_5 = p_3 = f_{13}p_1 + f_{23}p_2 \tag{2.1}$$

$$p_{10} = p_9 = f_{69}p_6 + f_{79}p_7 = f_{69}(f_{13}p_1 + f_{23}p_2) + f_{79}(f_{14}p_1 + f_{24}p_2). \tag{2.2}$$

Define

$$P = \begin{pmatrix} p_1 \\ p_2 \end{pmatrix}$$

and

$$M_1 = \begin{pmatrix} f_{13} & f_{23} \\ f_{13}f_{69} + f_{14}f_{79} & f_{23}f_{69} + f_{24}f_{79} \end{pmatrix}.$$

Then (2.1) and (2.2) can be written as

$$\begin{pmatrix} p_5 \\ p_{10} \end{pmatrix} = M_1 P.$$

Similarly, for the arcs 8 and 11 into $t_2$ we have, for some 2 by 2 matrix $M_2$,

$$\begin{pmatrix} p_8 \\ p_{11} \end{pmatrix} = M_2 P.$$

Destination $t_1$ can determine $P$ from $(p_5, p_{10})$ if and only if $M_1$ is non-singular, or equivalently, if the determinant $\det(M_1)$ is nonzero. Similarly, $t_2$ can determine $P$ from $(p_8, p_{11})$ if and only if $M_2$ is non-singular, or equivalently, if $\det(M_2)$ is nonzero. Each of these determinants is a multivariate polynomial in the variables $f_{ij}$. The problem is thus to determine values for the $f_{ij}$ variables that make $\det(M_1)$ and $\det(M_2)$ nonzero. It can be shown that, if $\det(M_1)\det(M_2)$ is not identically zero, then it is possible to find values of $f_{ij}$ such that $\det(M_1)\det(M_2) \neq 0$ whenever the size $q$ of the finite field $GF(q)$ exceeds the maximum degree of $\det(M_1)\det(M_2)$ with respect to any variable $f_{ij}$. Although the example of Figure 2.13 is for $k = 2$ (two incoming arcs at $s$), the results hold for any $k$ and for any number $D$ of destinations, so for $d = 1, 2, \cdots, D$ we generate a $k$ by $k$ matrix $M_d$ which is required to be non-singular.

One method, called *random network coding*, generates $\{f_{ij}\}$ by choosing each $f_{ij}$ from a uniform distribution over $GF(q)$ for a sufficiently large $q$. It can be shown that, with this method, the probability of obtaining a set $\{f_{ij}\}$ that makes each $M_d$ non-singular is $(1 - \frac{D}{q})^\beta$, where $\beta$ is the total number of $f_{ij}$ variables. We refer the reader to the books and articles surveyed in [100] for more details on how to determine the required $q$ and deterministic methods (including a polynomial time algorithm) for computing the $\{f_{ij}\}$. Also, bounds can be derived on the improvement of the information transmission rate using network coding, compared to not using it. The Avalanche file distribution protocol [71] has been implemented using random network coding.

## 2.6 Encodings of Multicast Trees

Consider a multicast group $g$ with only a few receivers. If the receivers are geographically dispersed over the network, $(\star, g)$ or $(s, g)$ state will be created on many nodes as flows travel to the receiver nodes. To avoid this, Arya, Turletti, and Kalyanaraman [7] in 2005 proposed encoding the multicast tree within every packet in a flow for $g$, thus completely eliminating creating $(\star, g)$ or $(s, g)$ state in the network for this flow. With this approach, multicast forwarding is accomplished by reading and processing packet headers. It also can reduce control overhead, while allowing real-time updating of the tree in case of a node or arc failure.

The multicast tree encoding must be of near minimal length, since it is included in each packet, and the processing of the encoding must be computationally simple. The representation can either specify $(i)$ only the receiver nodes (behind which are receiver hosts for $g$), or $(ii)$ the entire sequence of nodes in the tree; $(ii)$ can be useful for traffic engineering, to take a specified path.

For the *Link* $\star$ encoding described in [7], each arc has a unique index. The encoding has two components, $(i)$ a balanced parentheses representation of the tree, and $(ii)$ a preorder (depth first) list of the arc indices. The encoding is constructed by a preorder tree traversal: when an arc is visited for the first time, its arc index and a "(" are written, and when the arc is revisited after visiting all arcs in its subtree, a ")" is written. For a tree with $L$ arcs, this representation requires $2L$ parentheses and $L$ arc indices. Figure 2.15 shows the encoding for the full tree of Figure 2.14, and for the subtrees rooted at $Q$, $D$, and $F$; for clarity, the arcs represented by each encoding are also listed. In the forwarding of a packet, each node receives only the encoding of its subtree. A forwarding node reads the balanced parentheses once to identify its outgoing arcs and the encoding of subtrees to be forwarded to its children.

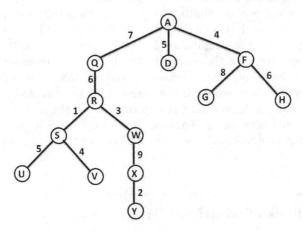

**Fig. 2.14** Tree encoding example

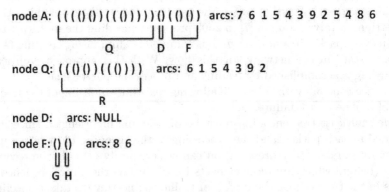

**Fig. 2.15** Multicast tree encoding

## 2.7 The Shape of Multicast Trees

In this section we present both empirical and theoretical results describing the shape of multicast trees. Several metrics can be used to describe a set of trees, including (i) *depth*, (ii) *degree frequency* (the percentage of nodes with degree $j$, for $j = 1, 2, \cdots$), and (iii) *average degree* (the expected value of the degree frequency distribution). One motivation for such studies is the recognition that, relative to unicast routing, there is increased complexity in designing and maintaining multicast networks, and that this complexity is typically justified by the savings in bandwidth, which is a function of the shape of the multicast tree. Another motivation is that, when designing a routing protocol, it is sometimes necessary to make assumptions about the topology of the networks over which the protocol will run.

We can study the topology of multicast trees in the Internet from two perspectives: (i) each node in the tree represents a router, and an arc in the tree represents a physical connection between two routers, or (ii) we map the actual multicast tree into an *AS-tree*, where each node in the AS-tree represents an AS containing one or more routers in the actual tree, and an arc $(i, j)$ exists in the AS-tree if there is at least one physical connection between a router in AS $i$ and a router in AS $j$ in the actual tree. Since multicast trees are dynamic, as receiver and source hosts appear and disappear, any empirical study is necessarily a snapshot in time.

In the early 2000s, there was considerable effort to characterize multicast trees observed in actual networks. This work is reviewed by Chalmers and Almeroth [20]. Many results showed the following power law relationship. Consider a tree with $R$ receiver hosts and $N$ nodes. Let $L_m$ be the total number of arcs in the tree. Suppose we find a unicast path between each pair of nodes in the tree; let $L_u$ be the total number of arcs in all these unicast paths. Let $\bar{L}_u = L_u/[(N)(N-1)/2]$, so $\bar{L}_u$ is the average path length, in hops, between any two nodes in the tree, Chuang and Sirbu [22] showed empirically that $L_m/\bar{L}_u = R^\alpha$ for some constant $\alpha$, where $0 < \alpha < 1$ (typically, $\alpha \approx 0.8$). This and other early studies ([85], [108]) used simulated networks, and assumed receivers were uniformly distributed in space, rather than using actual Internet data.

Define the efficiency $\delta$ of the multicast tree as $\delta = 1 - L_m/L_u$. As $\delta$ approaches 0, the cost of unicast and multicast are nearly equal, so there is little bandwidth savings from multicast. (A trivial example with $\delta = 0$ is a tree with a single arc.) As $\delta$ approaches 1, multicast becomes more efficient. A highly efficient multicast network is given in Figure 2.16, which shows 2 clusters connected by a single arc. Since $L_u$ is $O(K^2)$ and the multicast tree has $L_m = 2K + 1$ arcs, we have $1 - (L_m/L_u) \to 1$ as $K \to \infty$.

Following [20], using $L_m/\bar{L}_u = R^\alpha$ we write $L_m/\bar{L}_u = s_\varepsilon R^{\alpha-1}$, where $\bar{L}_u = s_\varepsilon L_u/R$, and where $s_\varepsilon$ is a scaling factor ($s_\varepsilon \approx 1$) introduced since the relationship between $\bar{L}_u$ and $L_u$ is not exact. This yields $\delta = 1 - s_\varepsilon R^\varepsilon$, where

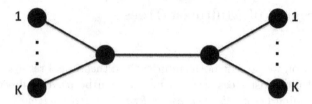

**Fig. 2.16** Multicast efficiency

$\varepsilon = \alpha - 1$, which provides a measure of multicast efficiency as a function of the number of receiver hosts. For $\alpha = 0.8$ we have $\varepsilon = -0.2$; the typical range for $\varepsilon$ is -0.38 to -0.27. Empirical results in [20] show that this relation holds under a wide range of behavior of receiver hosts. The formula implies that for groups with as few as 20 to 40 receiver hosts, multicast uses 55%-70% fewer arcs than unicast. A difference between [22] and [20] is that, if $J$ receiver hosts are connected to a leaf node, in [22] this counts as 1 receiver host, while in [20] this counts as $J$ receiver hosts. Another power law relationship given in [20] relates the number of $N$ of nodes in the tree to the number $R$ of receiver hosts: $N = s_\tau R^\tau$, where typically $\tau \in (0.66, 0.72)$.

Phillips, Sheker, and Tangmunarunkit [85], seeking to explain the power law $L_m/\bar{L}_u = R^\alpha$ observed by Chuang and Sirbu, assume the network is a $V$-ary tree (each node except the leaf nodes has $V$ descendants), of depth $D = \log_V N$. They showed that $L_u \approx R(D+1/\ln V - \ln R/\ln V)$, which is not a power law. The question was then taken up by Adjih, Georgiadis, Jacquet, and Szpankowski ([1], [2]). Define $a = R/N$. They observed that the results in [22] show a phase transition at $a = 1$: for $a \ll 1$ we have $L_m/L_u = \alpha R^\beta$ for some $\alpha > 0$ and some $\beta \approx 0.8$, and as $R \to \infty$ with $N$ fixed, the plot of $L_m/L_u$ flattens out. The first model studied in [2] is the same $V$-ary model used in [85]. For $1 \le k \le D$, there are a total of $V^k$ arcs connecting nodes at level $k-1$ to nodes at level $k$, and the probability that such an arc is in the tree after $R$ receiver hosts have been selected is $1 - (1 - V^{-k})^R$. The expected number of arcs in the tree is thus

$$L_m(R) = \sum_{k=1}^{D} V^k \left(1 - (1 - V^{-k})^R\right).$$

For $a << 1$, they show that

$$\frac{L_m(R)}{L_u} \approx R\left(1 + \frac{1-\gamma}{\ln N} + \frac{\ln V}{2\ln N} - \frac{\ln R}{\ln N} - \frac{\ln V}{\ln N}\psi(V, \ln a)\right),$$

where $\ln$ denotes natural logarithm, $\gamma$ is the Euler constant $0.571...$, and $|\psi(V, x)| < 0.153$ for all $x$ and $V < 1000$. For $a \to \infty$, they show that

$$\frac{L_m(R)}{L_u} = \frac{N}{D}\left(\frac{V}{V-1} - e^{-a}\right) - \frac{V}{V-1} - \frac{1}{2}\left(ae^{-a} + aVe^{-aV}\right) + O\left(\frac{1}{\ln R}\right).$$

The second model studied in [2] also assumes a $V$-ary tree of depth $D$, where all receivers are connected to leaf nodes. However in this model, the single arc between a node at level $k$ and the adjacent node at level $k-1$ is replaced by a concatenation of a random number of arcs, with no branching at intermediate nodes. Let $n_k$ be the average number of arcs inserted between levels $k$ and $k-1$, and assume that $n_k = \phi n_{k-1}$, where $0 \le \phi \le 1$. Define $\theta$ by $\phi = V^{-\theta}$. This is illustrated in Figure 2.17, for $\theta = 1$ and $D = 3$, where two nodes (smaller, not shaded) are inserted into each arc between levels 0 and 1, and one node is inserted into each arc between levels 1 and 2. With

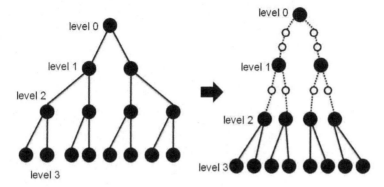

**Fig. 2.17** Self similar tree

this refinement, the average path length for a unicast connection is now

$$L_u^\theta = \sum_{k=1}^{D} V^{(D-k)\theta} = \frac{N^{\theta-1}}{V^\theta - 1},$$

and the expected number of arcs in the multicast tree is

$$L_m^\theta(R) = \sum_{k=1}^{D} V^{(D-k)\theta} V^k \left(1 - (1 - V^{-k})^R\right).$$

For $a \approx 0$, we obtain the power law

$$L_m^\theta(R)/L_u^\theta \approx c_1 R^{1-\theta} - c_2$$

for some constants $c_1$ and $c_2$, where $c_1$ and $c_2$ are independent of $R$ and $N$. As $a \to \infty$ with $N$ fixed, all leaf nodes belong to the multicast tree, so $L_m^\theta$ approaches $(NV^{1-\theta} - N^\theta V^{1-\theta})/(V^{1-\theta} - 1)$, which is the total number of links in the full multicast tree. Moreover, $L_m^\theta(R)/L_u^\theta$ shows a phase transition around

$a = 1$, in agreement with the empirical results of [22]. Although these results assumed receivers are attached only to leaf nodes, similar results hold when the model is extended to allow receivers to connect to non-leaf nodes that are branching nodes. Random trees are also studied in [16], which derives functional equations that can be used to estimate the advantage of multicast over unicast, based upon the tree branching distribution, the fraction of network nodes that are multicast enabled, and the number and location of receiver hosts.

# Chapter 3
# Dynamic Routing Methods

Multicast routing protocols generally include two components, one component for communication between routers and subtending hosts, and one component for communication between routers. Sections 3.1 and 3.2 discuss methods for communication between routers and hosts, and the subsequent sections of this chapter discuss methods for communication between routers.

Many Steiner tree methods, such as the centralized KMB method described in Section 2.1 above, are applicable to multicast routing only when the set of groups, and the sets of source and receiver nodes for each group, are known a priori, and these sets are static. The methods we describe in this chapter are applicable to multicast routing when the set of groups, and the sets of source and receiver nodes, are not known a priori, or these sets are dynamic. Some of these methods will use source trees, and others will use shared trees. The basic principle of tree-based dynamic routing methods (whether for a source tree or shared tree) is that a multicast stream should not be sent to a leaf node of a tree if there are no receiver hosts subtending the node.

For example, in Figure 3.1 (which is identical to Figure 2.6), the tree for group $g$ (indicated by the heavy solid lines) extends to leaf node 7, since there are receiver hosts behind node 7. Suppose both receiver hosts behind

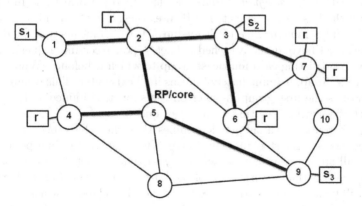

**Fig. 3.1** Shared tree

node 7 decide to leave $g$. Then node 7 notifies its upstream neighbor, which is node 3, that traffic for $g$ should not be sent to node 7. Node 3 then removes

35

arc $(3, 7)$ from its outgoing interface list (OIL) for group $g$. This is known as *pruning* node 7 from the tree. Suppose the receiver host behind node 6 also decides to leave $g$. Then node 6 informs node 3 that traffic for $g$ should not be sent to node 6, so arc $(3, 6)$ is deleted from the $(\star, g)$ OIL at node 3. Now node 3 is a leaf node of the tree. Since there are no receiver hosts behind node 3 itself, then 3 informs its upstream neighbor, node 2, that traffic for $g$ should not be sent to node 3, so node 2 deletes arc $(2, 3)$ from the $(\star, g)$ OIL at node 2. This process continues until no further pruning can be done.

Suppose now nodes 6, and 7, and 3 have been pruned from the tree for group $g$, and a host behind node 6 decides to rejoin $g$. First, the host informs node 6 that it is rejoining $g$. Then node 6 must rejoin the tree, so node 6 tells node 3 to send traffic for $g$ down to node 6, so node 3 adds arc $(3, 6)$ to its $(\star, g)$ OIL. Finally, node 3 tells node 2 to send traffic for $g$ down to node 3, so node 2 adds $(2, 3)$ to its $(\star, g)$ OIL.

## 3.1 Internet Group Management Protocol

Once a host learns about the existence of a multicast group (e.g., through an email or web site) it uses the *Internet Group Management Protocol* (IGMP) to announce its desire to join the group. A host wishing to join group $g$ sends an IGMP *membership report* to its local router. In turn, routers listen to the IGMP messages sent by hosts on a directly attached subnet, and periodically determine which multicast groups have at least one interested receiver host on the subnet. If there are multiple routers on a subnet, one of them, called the *designated router*, is elected to perform these functions. To query the hosts on a subnet, IGMP uses the *all-hosts* broadcast address 224.0.0.1 and a *Time to Live* (TTL) value of 1, which means that queries are not forwarded beyond the attached subnet. Upon receiving a query, a host sends a report listing each multicast group to which it belongs. When a host first joins a group, it immediately informs its local router. This ensures that, if the host is the first group member on the subnet, it immediately receives group traffic, without waiting for a router query.

When a router on a subnet determines that there have been no locally attached hosts interested in some group $g$ for a specified time period, the router will stop forwarding packets for $g$ to that subnet. The drawback of this approach is that, even if none of the hosts on a subnet want traffic for $g$, traffic will be forwarded until a specified period has elapsed, which is a waste of bandwidth and router resources. IGMP Version 2 provides a faster method to let a router know that it should stop forwarding traffic for $g$ to a subnet: a host may inform its local router, using an IGMP *leave group* message, that it does not want traffic for $g$. Upon hearing such a message, the router will initiate a query to determine if *any* host on the subnet wants traffic for $g$;

if no host does, then the router stops forwarding that traffic. Version 2 also allows a router to query for membership in a specific multicast group, rather than having attached hosts report on membership for all the groups to which they belong. Finally, IGMP Version 3 allows a host to inform its local router that it is interested in traffic for group $g$, but only for a specific set of sources for this group. If the host indicates interest in a group $g$ but does not specify a set of sources for $g$, traffic from any source for $g$ will be forwarded to the host. Similarly, IGMP Version 3 can allow a host to specify, for $g$, a set of sources from which it does *not* want to receive traffic. IGMP Version 3 is used with Source Specific Multicast (Section 3.7.4).

IGMP is only for IPv4; the IPv6 equivalent of IGMP is *Multicast Listener Discovery* (MLD) Version 2 (IETF RFC 3810).

## 3.2 IGMP Snooping

Suppose that, instead of a set of hosts being directly connected to a router, they instead connect to a Layer 2 switch (e.g., an Ethernet switch), and the Ethernet switch connects to the router. A Layer 2 switch receiving a multicast packet for group $g$ will, by default, send it to all attached hosts, even if these hosts have not signalled, using IGMP, their desire to receive packets for $g$. *IGMP snooping* at the switch can be used to stop the forwarding of packets to hosts uninterested in $g$.

With IGMP snooping, the switch examines the Layer 3 information in the IGMP packets exchanged between the hosts and the router. When the switch hears the IGMP membership report for $g$ from some host, the switch adds the host's port (i.e., interface) number to the switch's multicast table entry for $g$. Similarly, when the switch hears the IGMP leave group report for $g$ from some host, the switch deletes the host's port number from the switch's multicast table entry for $g$.

With IGMP Version 2, because IGMP membership report and leave group messages are transmitted as multicast packets, these messages cannot be distinguished from multicast data. Therefore the switch must examine every multicast data packet to see if it contains an IGMP membership report or leave group message. This may severely degrade the performance of the switch processor, especially for high bandwidth multicast data streams. To mitigate this, the IGMP checks can be performed using specially designed hardware, rather than using software or firmware. This issue disappears with IGMP Version 3, which sends IGMP reports to the fixed destination address 224.0.0.22, to assist Layer 2 switches in snooping.

## 3.3 Waxman's Method

A 1988 method of Waxman [110], which anticipated PIM join/prune methods, defines a greedy method for adding a node $x$ to a multipoint connection:
find a node $y$ that is already in the multipoint connection and whose distance $d(x, y)$ to $x$ is smallest; then connect $x$ and $y$ by a shortest path. An
alternative weighted greedy method relies on identifying a privileged node $z^\star$
which always remains in the multipoint connection; to add $x$, we find a node
$y$ in the connection which minimizes $f(x) = (1 - \alpha)d(x, y) + \alpha d(y, z^\star)$, where
$0 \le \alpha \le 1$, and connect $x$ and $y$ by a shortest path. When $\alpha = 0$, this is the
same as the greedy method. When $\alpha = 1/2$, the weighted greedy method is
equivalent to adding $x$ by a shortest path to $z^\star$. This use of $z^\star$ predates the
subsequent definition of a rendezvous point. Waxman's simulations showed
that $\alpha = 0.3$ achieves the best results. The worst case behavior is observed
when internal (non-leaf) nodes cease to be receiver nodes (i.e., if all subtending hosts have left $g$), but re-optimization of the tree is not allowed. This is
illustrated in Figure 3.2. Node $z^\star$ is the privileged node. First $a$ is added,

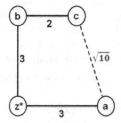

**Fig. 3.2** Greedy method

then $b$, and then $c$, and the cost of the greedy tree is 8. If we now delete $b$,
then the optimal tree consists of arcs $(z^\star, a)$ and $(a, c)$, with a cost of only
$3 + \sqrt{10} \approx 6.16$.

## 3.4 DVMRP

The first widely deployed multicast routing protocol is *Distance Vector
Multicast Routing Protocol* (DVMRP). It is based upon the RIP unicast routing protocol [68]. Each router running DVMRP maintains a list of adjacent
routers also running DVMRP. DVMRP is a distance vector method, which
means that adjacent nodes exchange routing updates (this is done every 60
seconds), and update their routing tables. The distance to a destination is
measured in hops, and the maximum allowed distance is 32 hops.

Suppose that DVMRP router $i$ is currently storing a distance of $c_i(s)$ to reach a certain source $s$ (if $i$ does not have $s$ in its routing table, then $c_i(s) = \infty$). If $i$ receives, from an adjacent router $j$, an update advertisement that $s$ is reachable from $j$ with cost $c_j(s)$, and if $c_j(s) + 1 < c_i(s)$, then $i$ now stores the new cost $c_j(s) + 1$ to reach $s$ (1 is added since an extra hop is incurred), and records that the best path to $s$ is via node $j$. Then node $i$ advertises back to $j$ that $c_i(s) = \infty$. This *poison reverse* message informs $j$ that $i$ is downstream of $j$ in the tree rooted at source $s$, so $j$ adds the interface to $i$ to its OIL for this tree.

DVMRP builds a source tree for each $(s, g)$. When a node receives the first packet for $(s, g)$, it performs an RPF check. If the check passes, the node creates $(s, g)$ state and creates an $(s, g)$ OIL, based upon the poison reverse messages it has received. In a stable network, where all routing updates have converged, the RPF check is not needed, since a node's OIL will point only to downstream neighbors who sent a poison reverse. However, the RPF check ensures correct, loop-free behavior in the event that the routing tables have not yet stabilized.

Initially, $(s, g)$ packets are flooded to all DVMRP routers, whether or not there are interested hosts for $(s, g)$ behind the routers. This results in $(s, g)$ state being created on all the DVMRP routers. If there are no interested hosts behind a leaf node $n$, it sends an $(s, g)$ *prune* message towards the source. The upstream node $y$ receiving this prune on interface $i$ removes $i$ from its $(s, g)$ OIL. Node $y$ may in turn send an $(s, g)$ prune if it is now a leaf node without any interested subtending hosts. With DVMRP, pruning does not mean that $(s, g)$ state is removed from a pruned node; rather, it means only that $(s, g)$ traffic is not sent to the pruned node. The $(s, g)$ state continues to exist in case the pruned node needs to *graft* (i.e., re-attach) itself to the $(s, g)$ tree; this occurs if a host behind the node now wants to receive $(s, g)$ packets. With DVMRP, each prune times out after a fixed interval, after which $(s, g)$ traffic is again flooded. So a pruned node, in order to stay pruned, must periodically issue prune requests.

If a pruned node $n$ wishes to re-join the $(s, g)$ tree, e.g., because of an interested subtending host, it issues a graft message to its upstream neighbor $y$ on the shortest path to $s$; node $y$ then re-adds, to its $(s, g)$ OIL, the interface on which the graft was received. If node $y$ itself was pruned, it issues a graft message to its upstream neighbor, and this continues until the graft reaches a non-pruned node on the $(s, g)$ tree.

DVMRP is not currently utilized in large real world production networks, because its 32 hop limit makes it unsuitable for large networks, because its distance vector routing updates may be slow to converge, and because its flood and prune approach is more wasteful of bandwidth and state than PIM-based join/prune approaches.

## 3.5 Multicast OSPF

*Multicast Extensions to OSPF* (MOSPF) is a multicast protocol designed
by Moy, the author of OSPF ([75], [76]); Moy indicates that MOSPF is based
on 1998 work of Steve Deering. The first MOSPF implementation for general
use was released in 1992. We begin with a brief description of OSPF. OSPF
is a link state protocol, which means that each node stores a view of the
network topology and computes shortest paths based on this view. An OSPF
domain can be either (*i*) a *flat* network, in which all nodes belong to one area,
called *area 0*, or (*ii*) a *multi-area* network, in which nodes are partitioned into
multiple areas, with the restriction that there is always an area 0 together
with at least one *non-zero* area (see Figure 3.3 ). In multi-area OSPF, while

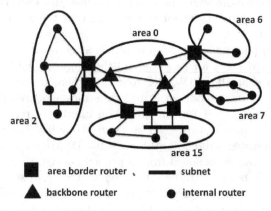

**Fig. 3.3** OSPF areas

each node (specifically, a *loopback* address for the node) is assigned to a
unique area, the different interfaces of a given node might belong to different
non-zero areas, in which case the node is an *area border router* (ABR). Each
ABR must have at least one interface in area 0. If all the interfaces of a
node belong to area 0, the node is a *backbone* router; if all the interfaces
of a node belong to the same non-zero area, the node is an *internal router*.
Since the complexity of computing a shortest path in a network with $N$ nodes
is typically $O(N \log N)$ or $O(N^2)$, depending on the implementation of the
shortest path method, multi-area OSPF is used when the number of nodes
in the network exceeds the limit each node can store or process.

Each OSPF node in an area floods *link state advertisements* (LSAs) to
all other nodes in the area. There are many types of LSAs; for brevity we
mention only four. A *router* LSA is originated by each OSPF router, and
specifies the router's IP address and all its interfaces; for each interface, the
neighboring node IP address and the cost to that neighbor (for use in shortest

path calculations) is specified. A *network* LSA is used to advertise a subnet and the routers attached to the subnet; the network LSA is originated by the designated router for the subnet. For multi-area OSPF domains, router and network LSAs originated in some area $A$ are only flooded within $A$ but not to other areas.

For multi-areas domains, a *summary* LSA is sent by an ABR attached to non-zero area $A$ to summarize a set of reachable addresses (routers and subnets) in $A$ and a cost to reach the summary address from the ABR. More than one summary LSA might be configured to summarize the reachable addresses in $A$. All summary LSAs for $A$ are injected by the ABR into area 0, which floods them within area 0, and hence makes them available to all other ABRs. An ABR not attached to $A$ will in general will receive multiple summary LSAs (with different costs) for a given summary address for $A$, so the ABR ($i$) chooses the minimal cost, ($ii$) creates a routing table entry for this summary address with this minimal cost, and ($iii$) re-advertises this summary address, with this minimal cost, into its attached non-zero areas by originating its own summary LSA.

The last LSA type we mention is specifically for MOSPF. A node $n$ in OSPF area $A$, having received IGMP membership reports from subtending hosts interested in group $g$, creates a *group membership* LSA indicating that $n$ wants to receive packets for $g$. For a subnet, the designated router issues a group membership LSA for $g$ if there are hosts on the subnet interested in $g$. A group membership LSA for area $A$ is flooded within $A$, but not to other areas; thus all nodes in $A$ have the same view of which nodes want to receive which groups.

Consider first the case of a flat (single area) OSPF domain. When the source node $n(s)$ receives the first $(s, g)$ packet from a subtending host $s$, the source node calculates, using the router and network LSAs, a shortest path tree with itself as the root and spanning all the routers and subnets in the domain. If there are multiple such shortest path trees, a tie breaker (known to all nodes in the domain) is utilized. Then the group membership LSAs are used by $n(s)$ to prune branches not leading to a receiver node for $g$. Each $(s, g)$ packet is sent out by $n(s)$ on all interfaces in its $(s, g)$ OIL. When the first $(s, g)$ packet is received by a downstream node $x$, then $x$ itself calculates a shortest path $(s, g)$ tree routed at $n(s)$, and prunes the tree using the group membership LSAs. Nodes downstream of $x$ themselves perform this on-demand calculation upon receiving the first $(s, g)$ packet, and so on. Thus each router receiving an $(s, g)$ packet calculates the same shortest path $(s, g)$ tree and prunes the tree. This is a very different mechanism than PIM based methods (Section 3.7) in which a receiver node sends an explicit join to attach to a tree.

Now consider the case of multi-area OSPF, where in general all receiver nodes for a group $g$ are not contained within a single area. To support inter-area multicast routing, MOSPF allows an ABR to be designed as an *inter-area multicast forwarder* (IAMF). Consider an IAMF with one or more interfaces in some non-zero area $A$. All multicast packets generated within $A$ are sent to the IAMF; this is accomplished by a feature in MOSPF that allows configuring a node as a *wild card multicast receiver* to which all multicast traffic, for any $s$ and $g$, should be forwarded. The IAMF summarizes the group membership for $A$, by determining which multicast groups have been joined by hosts in $A$. The IAMF then injects into the backbone area a group membership LSA for each group with members in $A$. (However, the backbone does not itself inject summary group information, whether for the backbone or for other non-zero areas, into $A$.) This ensures that the backbone has knowledge of group membership for the entire OSPF domain, and thus can forward $(s, g)$ multicast packets sourced in $A$ to other non-zero areas containing receiver nodes for $g$. We consider two possibilities.

First, suppose an $(s, g)$ multicast packet is received by a node $n$ attached to non-zero area $A$ and the source node for the packet is also in $A$. Then $n$ has sufficient information to build an $(s, g)$ shortest path tree rooted at the source node. The $(s, g)$ OIL at $n$ is used to determine on which interfaces a replica of the packet should be sent. Since $n$ does not know which other areas contain receiver nodes for $g$, an interface in the $(s, g)$ OIL at $n$ can be pruned, using the group membership LSAs, only if the interface does not lead to either a receiver node for $g$ in $A$ or to a wild card multicast receiver.

Now suppose an $(s, g)$ multicast packet is received by a node $n$ attached to non-zero area $A$, and the source node for the packet is not in $A$. Then $n$ does not know the detailed topology of the source OSPF area, but $n$ does possess summary LSAs providing the approximate cost from the IAMFs of $A$ to the source node. Hence $n$ can calculate an $(s, g)$ tree whose root is the source node; the arcs incident to the root in this tree represent paths from the root to the IAMFs attached to $A$. Since $n$ also knows the full topology of $A$, this tree also spans the nodes of $A$ wishing to receive packets for $g$. If $n$ is attached to multiple non-zero areas, this procedure will cause $n$ to build multiple $(s, g)$ trees. In this case, MOSPF merges these multiple trees into a single $(s, g)$ tree, and the individual trees are discarded.

There is considerable overhead with MOSPF, since group membership LSAs are flooded within each area and are also injected into the backbone, since a shortest path source tree is generated for each $(s, g)$, and since each node receiving an $(s, g)$ packet must calculate an $(s, g)$ tree. For these reasons, it is generally recognized that MOSPF cannot support the scale required for Internet multicast [84].

## 3.6 Core Based Trees

*Core Based Trees* (CBTs) were proposed in 1993 by Ballardie, Francis, and Crowcroft [10], and extended in 1998 by Ballardie, Cain, and Zhang [9].

In a CBT, a single core node is assigned for each $g$, and a $(\star, g)$ shared tree, rooted at the core node, is built. The core could be the same for all groups, or different cores could be used for different groups. The identity of the core node for the $(\star, g)$ tree is communicated to all routers in the multicast domain using the PIM Bootstrap procedure described in Section 3.8.3 below. A CBT is bidirectional: packets for $g$ can flow up or down the tree. A node $n$ starts the process of joining the tree when it receives an IGMP membership report from a subtending host who wants to receive $g$; node $n$ then sends a $(\star, g)$ join request to the next hop router on the shortest path to the core node. Since this next hop router is determined using the unicast routing table, CBT is, like PIM, protocol independent, meaning it will work with any unicast routing protocol. The join is forwarded hop by hop until either the core is reached, or a node on the $(\star, g)$ tree is reached; once this occurs the branch of the shared tree is established, and packets can flow up or down the branch. The $(\star, g)$ *forwarding information base* (FIB) stored at each node visited by these joins contains ($i$) the group $g$, ($ii$) the RPF interface (the interface corresponding to a shortest path to the core), and ($iii$) the $(\star, g)$ OIL (the child interface list). The RPF neighbor for the core node is the core itself.

Suppose for some node $n$ there is a subtending source host $s$ for $g$ but no subtending receiver host for $g$. Then $n$ is not a receiver node for $g$, and hence will not join the shared tree for $g$. To allow traffic from $s$ to reach receivers for $g$, with CBT Version 2, node $n$ encapsulates packets from $s$ in a bidirectional IP-within-IP tunnel, where the ends of the tunnel are $n$ and the core. When the encapsulated packets reach the core, they are decapsulated and forwarded down the tree to interested receivers. CBT Version 2 provides no way to inform $n$ to stop sending packets from $s$ to the shared tree for $g$, even when no receivers have joined $g$. Also, since the tunnel is bidirectional, packets for $g$ will flow from the core to $n$, even though $n$ is not a receiver node. With CBT Version 3, a unidirectional path, not requiring encapsulation, is established from $n$ to the core. This path allows packets from $s$ to reach the shared tree, but does not allow traffic to flow down to $n$.

For subnets, CBT supports the election of a *designated router* which determines the best path to the core (since different routers on the subnet might have inconsistent routing tables, leading to loops). When the first host on a subnet indicates it wants to join $g$, the designated router issues the upstream join message.

Forwarding on a CBT works as follows. When a node $n$ on the tree receives a packet for $g$, it discards it if there is no $(\star, g)$ FIB, or if the packet was received on an interface not in the $(\star, g)$ FIB. Otherwise, $n$ forwards the packet out over all interfaces in the $(\star, g)$ FIB, except for the interface on which the packet was received. No RPF check is used in CBT; the check is not applicable, since packets flow both up and down the shared tree. The lack of an RPF check requires ensuring, e.g., through network management, that no routing loops are created.

Periodic keepalive messages are exchanged by each non-core node $n$ on the tree and its RPF neighbor to ensure that $n$ still needs to receive packets for $g$. The tree can be pruned when a leaf node $n$ with no interested subtending hosts sends an upstream prune message. If the arc between $n$ and the RPF neighbor $p$ is a point-to-point arc, then $p$ simply deletes $(p, n)$ from its $(\star, g)$ OIL. However, $p$ might also receive a prune message on a child interface $i$ that is a subnet. If $p$ deletes $i$ from its $(\star, g)$ OIL, then other interested hosts on this subnet would no longer be able to receive packets for $g$. To prevent this from happening, $p$ waits a specified time period before deleting $i$ from its $(\star, g)$ OIL. This gives other routers on the subnet the opportunity to override the prune by sending a join request; any join request received by $p$ on interface $i$ during this period cancels the prune.

The core is a single point of failure. If it fails, the core tree may be partitioned into a set of disjoint trees. To ensure that each of these disjoint trees can operate as a separate CBT, the CBT protocol allows additional nodes to be specified as backup cores. Each router that can serve as a core gets a priority value. Each node knows the priority values of the nodes that are core candidates, and a join is sent to the (unicast) address of the reachable candidate node with the highest priority. With this approach, if the core fails, a set of disjoint CBTs will be created, with each non-core node belonging to exactly one CBT.

It is also possible to have multiple active cores for a single CBT. This may be desirable if the receivers span a wide geographic area. The set of active core nodes must be interconnected. All the active cores have the same priority, and non-core nodes send packets to the nearest active core. Although this is feasible, due to the complexity of managing multi-core trees, the recommendation in [10] is to only use a single core. (The MSDP protocol discussed in Section 5.2 does allow multiple PIM rendezvous points to communicate.) In practice, the CBT protocol was not implemented beyond experimental networks, in large part because the focus shifted to creating shortest path trees, which PIM provides.

## 3.7 Protocol Independent Multicast

*Protocol Independent Multicast* (PIM) refers to a set of multicast routing methods, all based upon the idea that unicast routing is the foundation of multicast routing, and multicast routing should be able to utilize any unicast routing protocol, e.g., ISIS or OSPF. All PIM routing methods utilize the Reverse Path Forwarding (RPF) check, and the RPF check utilizes the unicast forwarding table. The PIM protocol itself is not responsible for unicast routing updates between routers; that is the responsibility of the underlying unicast routing protocol. Rather, PIM is used for join, prune, and source register messages.

We will review four PIM multicast routing methods: ($i$) PIM Dense Mode, ($ii$) PIM Sparse Mode, ($iii$) Bi-directional PIM, and ($iv$) Source Specific Multicast (SSM). All four methods rely on PIM neighbor adjacencies, which are created when a node sends a *PIM hello* out on each of its multicast enabled interfaces. A PIM adjacency must be periodically refreshed with a Hello, and an adjacency expires if a Hello is not received by a PIM neighbor after some configured time period.

## 3.7.1 PIM Dense Mode

PIM Dense Mode was developed in 1998 by Deering, Estrin, Farinacci, Jacobson, Helmy, Meyer, and Wei [34]. This method is an example of a *push* method [112], which assumes that each node and subnet contains at least one host interested in receiving the multicast stream. PIM Dense Mode uses source trees. Consider a given group $g$ and a source host $s$, subtending the source node $n(s)$, and sending to $g$. The method generates an $(s, g)$ *broadcast tree*, rooted at $n(s)$ and spanning the entire network, and $(s, g)$ packets are flooded over this tree. In this broadcast tree, each node sends $(s, g)$ packets to all its downstream PIM neighbors, which in general might cause a given node to receive the same packet from two upstream neighbors.

If two or more routers on a common subnet receive an $(s, g)$ packet from upstream neighbors, we want only one router to forward the packet onto the subnet. This is accomplished using an *assert* mechanism, which selects the router on the subnet with the smallest unicast cost to the source node. All routers on the subnet other than the assert winner prune from their $(s, g)$ OIL the interface to the subnet.

A node $n$ prunes itself from the $(s, g)$ broadcast tree if any of these four conditions holds: ($i$) if $(s, g)$ packets arrive on a point-to-point interface other than the RPF interface, ($ii$) if $n$ is a leaf node and $n$ is not a receiver node, ($iii$) if $n$ is not a receiver node and its $(s, g)$ OIL is empty, and ($iv$) if $n$ is on a subnet, $n$ is not a receiver node, $n$ has received a prune message from some node on the subnet, and no other node on the subnet has issued an $(s, g)$ join. However, pruned nodes can graft themselves back onto the tree upon receiving a join from an interested subtending host.

In the original PIM Dense Mode protocol specification, PIM prunes time out after 3 minutes, after which a new broadcast tree is created and pruning starts again. This periodic re-flooding of $(s, g)$ packets to non-receiver nodes is not efficient for high bandwidth streams. In more recent PIM Dense Mode implementations, the source node $n(s)$ periodically sends *state-refresh* messages down the original broadcast tree as long as $s$ is still sending packets to $g$. These state-refresh message reset the PIM timers, so the prunes do not time out [111].

Each node in the $(s, g)$ broadcast tree maintains $(s, g)$ state. If for this $g$ there is another source $\bar{s}$, also behind $n$, then each node in the network also maintains $(\bar{s}, g)$ state.

Even though PIM Dense Mode uses the same flood-and-prune approach as DVMRP, PIM Dense Mode does not have the burden of maintaining unicast routing information, so it is more "lightweight" than DVMRP. PIM Dense Mode is efficient when most nodes in the domain are receiver nodes. If few nodes are receiver nodes, since PIM Dense Mode initially sends packets on the broadcast tree to every node, this method consumes more bandwidth on arcs, and creates state on more nodes, than methods which send traffic only where it is wanted. For this reason PIM Dense Mode is not used in large scale multicast networks.

### 3.7.2 PIM Sparse Mode

*PIM Sparse Mode* (PIM-SM) is a popular multicast protocol, the most popular as of 2004 [89]. The original 1997 specification for PIM-SM was developed by Estrin, Farinacci, Helmy, Thaler, Deering, Handley, Jacobson, Liu, Sharma, and Wei [37]; the most recent specification, issued in 2006, is RFC 4601 [41]. PIM-SM is a *join* (i.e., *pull*) method, rather than a flood and prune (i.e., *push*) method, since hosts signal their interest in receiving a multicast stream. For each $g$, PIM-SM builds a shared tree, rooted at a *rendezvous point* (RP). The RP serves the same function as the core node in a CBT. The RP must be made known to each node in the network; methods for this are discussed in Section 3.8 below. Packets flow *down* the shared tree from the RP to each receiver node.

Packets cannot flow up the tree from a source host to the RP, since the RPF check at a node $n$ only accepts packets coming from the RPF neighbor of $n$ (the neighbor of $n$ on the shortest path to the RP). A packet coming from a node downstream of $n$ would fail the RPF check and be discarded. Therefore, with PIM-SM, to get packets from the source host to the RP, a *source path* is built from the source node towards the RP (just as CBT builds a path from a source only node to the core node). The source path is computed using unicast routing, with the source node as the unicast source, and the RP as the unicast destination. For example, in Figure 3.1 above, suppose the shortest path from node 1 to node 5 uses arcs $(1, 2)$ and $(2, 5)$. Then the source path for $g$ consists of these two arcs. Even though these two arcs are also part of the shared tree for $g$, the source path and the shared tree are logically distinct.

Each node on the source path for a source host $s_1$ and group $g$ gets $(s_1, g)$ state. Thus, in Figure 3.1, we have $(s_1, g)$ state on nodes 1, 2, and 5. The flow of traffic from $s_1$ to the receivers is from $s_1$ towards the RP, and then down from the RP to the receivers. The exception to this rule is if the receiver node is located on the source path between the source and the RP, in which case packets flow from the source to the receiver, without reaching the RP. For example, if there were a receiver behind node 2, the traffic from $s_1$ would reach this receiver by traversing the single arc $(1, 2)$, without having to first reach the RP. Suppose we have a second source host $s_2$ behind node 3, and the shortest path from $s_2$ to the RP uses arcs $(3, 2)$ and $(2, 5)$. Then these two arcs constitute the source path from $s_2$ to the RP, and $(s_2, g)$ state will appear on the nodes 3, 2, and 5.

As already mentioned, the source path and shared tree may share some arcs in common. For example, consider Figure 3.4. The shared tree consists

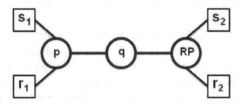

**Fig. 3.4** Bidirectional traffic on an arc

of the two arcs $(RP, q)$ and $(q, p)$. Traffic from $s_2$ to $r_1$ will take the path $s_2 \rightarrow RP \rightarrow q \rightarrow p \rightarrow r_1$. The source path from $s_1$ consists of the arcs $(p, q)$ and $(q, RP)$. Traffic from $s_1$ to $r_2$ will take the path $s_1 \rightarrow p \rightarrow q \rightarrow RP \rightarrow r_2$. Thus arc $(q, p)$ is in the shared tree, and arc $(p, q)$ is in the source path from $s_1$. However, packets from $s_1$ to $r_1$ will not travel to the RP, and instead take the path $s_1 \rightarrow p \rightarrow r_1$; in this case, node $p$ is a *turnaround* node.

## Shortest Path Trees.

Although PIM-SM always builds a shared tree for each $g$, an important feature of PIM-SM is the ability for a receiver node to receive traffic over a *Shortest Path Tree* (SPT) rooted at the source node, rather than over the shared tree. This option is desirable if latency (delay) is a primary concern. A receiver node $x$ wanting to receive packets for $g$ over an SPT must first learn the identity of (i.e., the IP address of) the source host $s$. However, $x$ does not initially know this address, but rather knows only the group address $g$. The way that PIM-SM provides to $x$ the address of $s$ is to first build the shared tree for $g$. When $s$ begins sending packets, they traverse the shared tree, finally reaching $x$, at which time $x$ will know about $s$.

Once $x$ knows $s$, it creates $(s, g)$ state and sends a PIM join towards $s$. The interface used by $x$ for this join is the RPF interface towards $s$, as computed by the unicast routing tables. The join sent by $x$ follows the shortest path towards $s$; if the next node on the path is $y$, then $(s, g)$ state is created at $y$, and the interface leading to $x$ is added to the $(s, g)$ OIL at $y$; now $y$ issues a join along its RPF interface towards $s$, and this continues until we reach the source node $n(s)$ which $s$ subtends. The SPT rooted at $n(s)$ for $g$, which we denote by $SPT(s, g)$ is, at this point, the path from $n(s)$ to $x$ established by these joins.

Now suppose another receiver node $x'$ also wants to receive packets from $s$ over an SPT. Once $x'$ also learns $s$, it too can send a join towards $s$. A series of joins again propagates towards $s$, where $(s, g)$ state is created and the $(s, g)$ OIL is updated at each node visited by these joins. Ultimately, these joins reach $n(s)$ or some other node $z$ on $SPT(s, g)$. At this point $n(s)$ or $z$ adds, to its $(s, g)$ OIL, the interface on which the join was received, and a new branch of $SPT(s, g)$ has been created. An RPF check is used with all SPTs.

The shared tree for $g$, which spans all receiver nodes of $g$, does not disappear once a shortest path tree is created. Rather, the shared tree continues to exist and span *all* receiver nodes of $g$. However, suppose that for some given $s$ and $g$ a receiver node $x$ has begun receiving $(s, g)$ packets over $SPT(s, g)$. Then $x$ no longer wants to receive $(s, g)$ packets over the shared $(\star, g)$ tree. Assume $x$ is a leaf node of the shared tree, and suppose the RPF neighbor for the $(s, g)$ OIL at $x$ differs from the RPF neighbor for the $(\star, g)$ OIL at $x$ (which means that $x$ receives $(s, g)$ packets on two different interfaces, one on the shortest path to $s$ and the other on the shortest path to the RP). Then $x$ sends an $(s, g)$ *RP-bit Prune* message towards the RP. Let $y$ be the next node on the path along the shared tree from $x$ to the RP. If $y$ is not on $SPT(s, g)$, then when $y$ receives the prune message, say on interface $i$, it updates its $(s, g)$ OIL so that $(s, g)$ packets are no longer sent out to $x$ on interface $i$. If $y$ is on $SPT(s, g)$, then it needs to continue sending $(s, g)$ packets out on $i$. If there are no other outgoing interfaces at $y$ over which $(s, g)$ packets are sent, and if $y$ is not the RP, then $y$ sends an RP-bit Prune message towards the RP. If this $(s, g)$ RP-bit Prune message reaches the RP, if the RP is not on $SPT(s, g)$, and if the RP itself now has no other interfaces over which $(s, g)$ packets should be forwarded, then the RP can send an $(s, g)$ RP-bit Prune message towards the source node.

This is illustrated in Figure 3.5. Assume that for $g$ there is only one source host $s$, behind node $t$, and only one receiver host $r$, behind node $x$. The shared tree consists of the two arcs $(RP, y)$ and $(y, x)$. The shortest path tree $SPT(s, g)$ consists of the single arc $(t, x)$. Once $x$ receives $(s, g)$ packets over $SPT(s, g)$, it sends (on interface $i_1$) an $(s, g)$ RP-bit Prune to $y$, which updates its $(s, g)$ OIL to not send $(s, g)$ packets out of interface $i_2$. Since $y$ now has no need for $(s, g)$ packets, it sends (on interface $i_3$) an $(s, g)$ RP-bit Prune to $RP$, which updates its $(s, g)$ OIL to not send $(s, g)$ packets out of

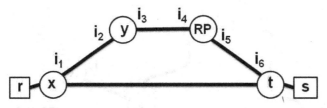

**Fig. 3.5** (s,g) RP-bit Prune

interface $i_4$. Finally, since $RP$ now has no need for $(s, g)$ packets, it sends (on interface $i_5$) an $(s, g)$ RP-bit Prune to $t$, which updates its $(s, g)$ OIL to not send $(s, g)$ packets out of interface $i_6$.

There will be a brief period of time when $x$, and hence the receiver host, receive duplicate $(s, g)$ packets: one copy via the shared tree, and one copy via $SPT(s, g)$. The multicast application running at the receiver host must be tolerant of receiving multiple copies of packets. Once $x$ has pruned itself from the shared tree, it will receive only a single copy of each $(s, g)$ packet.

Prior to receiving the $(s, g)$ RP-bit Prune from $x$, the $(\star, g)$ OIL at node $y$ has only one entry, specifying that $(\star, g)$ packets, if received on interface $i_3$, should be forwarded out of $i_2$. When $y$ receives the $(s, g)$ RP-bit Prune from $x$, it cannot delete the $(\star, g)$ entry, since then packets for $g$ from other sources would not be forwarded. Instead, $y$ stores the information that $(s, g)$ packets (for this particular $s$) arriving on $i_3$ should not be forwarded out of $i_2$. Thus this type of prune does not delete an OIL entry, but rather indicates an exception to the $(\star, g)$ OIL entry.

The disadvantage of using an SPT is that $(s, g)$ state is created on each node on $SPT(s, g)$. Under the assumption that latency is less important for low bandwidth streams, one way, used in practice, of limiting the creation of SPTs is to create an SPT only for a source that is transmitting at a sufficiently high rate. Once a receiver node on the shared tree for $g$ sees that the $(s, g)$ flow rate (i.e., bandwidth) flowing down the $(\star, g)$ tree exceeds some threshold, it starts the process of sending a join towards $s$. Once the receiver node has created $(s, g)$ state and become part of $SPT(s, g)$, the node monitors the $(s, g)$ flow rate on $SPT(s, g)$. If this rate falls below a configured threshold, the node reverts to using the shared $(\star, g)$ tree, and prunes itself from $SPT(s, g)$.

Since PIM-SM is so widely deployed, we provide more detail on how it works, using Figure 3.6. Additional details can be found in [112].

**Procedure:** host $rcvr_1$ joins the shared tree:

1. $rcvr_1$ uses IGMP to join $g$ (assume $rcvr_1$ is the first host subtending $R_1$ to join $g$).
2. $R_1$ creates $(\star, g)$ state and places $i_1$ in its $(\star, g)$ OIL.

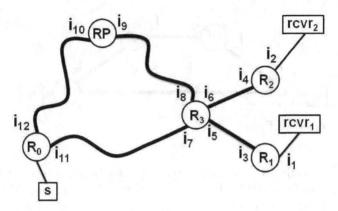

**Fig. 3.6** PIM sparse mode

3. Since $R_1$ had to create $(\star, g)$ state, it sends a $(\star, g)$ join along the best unicast path to $RP$ to join the shared tree; this path leaves $R_1$ on interface $i_3$.
4. The join reaches $R_3$ on interface $i_5$. Since $R_3$ had no $(\star, g)$ state, then $R_3$ creates $(\star, g)$ state, places $i_5$ in its $(\star, g)$ OIL, and forwards the join along the best unicast path to $RP$; this path leaves $R_3$ on interface $i_8$.
5. Each subsequent node on the shortest path from $R_3$ to $RP$ in turn sends a $(\star, g)$ join towards $RP$, until finally a join reaches $RP$ on incoming interface $i_9$, so $RP$ places $i_9$ in its $(\star, g)$ OIL. The shared tree from $RP$ to $R_1$ is now established.

Note that receivers can join a shared tree even if there is no active source. Also, the forwarding path from $RP$ to $R_1$ is the shortest path from $R_1$ to $RP$ taken by the PIM joins. This path from $RP$ to $R_1$ will be used by all packets flowing down the shared tree from $RP$ to $R_1$ (packets always flow in the opposite direction as the PIM join). If all arc costs are symmetric $(c_{ij} = c_{ji})$, then this path is also a shortest path from $RP$ to $R_1$; if the costs are asymmetric, this path may not be a shortest path from $RP$ to $R_1$.

**Procedure:** host $rcvr_2$ joins the shared tree:

1. $rcvr_2$ uses IGMP to join $g$.
2. $R_2$ creates $(\star, g)$ state, adds $i_2$ to its $(\star, g)$ OIL, and sends a $(\star, g)$ join along the best unicast path to $RP$; this path leaves $R_2$ on interface $i_4$.
3. The join reaches $R_3$ on interface $i_6$. Since $R_3$ already has $(\star, g)$ state, $R_3$ adds $i_6$ to its $(\star, g)$ OIL. There is no need for $R_3$ to send another join.

**Procedure:** host $rcvr_2$ leaves the shared tree:

1. $rcvr_2$ sends an IGMP leave group message to $R_2$.
2. $R_2$ receives this message on $i_2$, and removes $i_2$ from its $(\star, g)$ OIL. Since this OIL is now empty, $R_2$ sends a $(\star, g)$ prune towards $RP$; this prune leaves $R_2$ on interface $i_4$.

3. The prune is received by $R_3$ on interface $i_6$, so $R_3$ removes $i_6$ from its $(\star, g)$ OIL. Since the $(\star, g)$ OIL at $R_3$ is not empty, $R_3$ does not send a prune message towards $RP$.

**Procedure**: source host $s$ wants to send traffic to group $g$:

1. Source host $s$ begins to send packets addressed to $g$. (The source host does not first signal $R_0$ using IGMP; it simply starts transmitting.)

2. As $s$ begins to send, router $R_0$, after verifying that $g$ is in the proper multicast address range, creates $(s, g)$ state, and encapsulates each packet in a *PIM register* message, which is unicasted to $RP$.

3. $RP$, upon receiving a PIM register packet, decapsulates it. If $RP$ has an empty $(\star, g)$ OIL, then $RP$ discards the packet, and sends an $(s, g)$ PIM *register-stop* message to $R_0$, which tells $R_0$ to stop encapsulating $(s, g)$ packets in PIM register messages (since there are no interested receiver hosts). If $RP$ has a nonempty $(\star, g)$ OIL, then $(i)$ $RP$ forwards the packet down the shared tree (which, in this example, means that $RP$ sends the packet out of interface $i_9$ only), and $(ii)$ $RP$ sends an $(s, g)$ join towards $s$, along the best unicast path from $RP$ to $R_0$; this path leaves $RP$ on interface $i_{10}$.

4. At each node on the path from $RP$ to $R_0$, $(s, g)$ state is created, the incoming interface (on which the $(s, g)$ join was received) is added to the $(s, g)$ OIL, and a join is sent towards $R_0$.

5. When the last join reaches $R_0$, on interface $i_{12}$, this interface is added to the $(s, g)$ OIL at $R_0$. The source host $s$ can now send native (i.e., non-encapsulated) packets to $RP$ along this path. Now the source host $s$ is sending both encapsulated and non-encapsulated packets to $RP$. When $RP$ receives the first non-encapsulated $(s, g)$ packet, $R_0$ no longer needs to send $(s, g)$ packets using encapsulated unicast, so $RP$ sends an $(s, g)$ PIM register-stop message to $R_0$.

Note that the forwarding path from $R_0$ to $RP$ is the shortest path from $RP$ to $R_0$ taken by the PIM joins. This path will be taken by all packets sourced by $s$ (packets always flow in the opposite direction as the PIM join). If all arc costs are symmetric, then this path is also a shortest path from $R_0$ to $RP$.

**Procedure**: switch to the shortest path tree for the $(s, g)$ stream:

1. The bandwidth of the $(s, g)$ stream received at $R_1$ exceeds a specified *SPT-threshold*, which triggers $R_1$ to create $(s, g)$ state, add $i_1$ to its $(s, g)$ OIL, and send an $(s, g)$ join towards $s$. This join is sent on the best unicast path from $R_1$ to $s$, and leaves $R_1$ on interface $i_3$.

2. $R_3$ receives the $(s, g)$ join on interface $i_5$, creates $(s, g)$ state, and adds $i_5$ to its $(s, g)$ OIL. Since $R_3$ previously had no $(s, g)$ state it now sends an $(s, g)$ join towards $s$ on the best unicast path from $R_3$ to $s$; this path leaves $R_3$ on interface $i_7$.

3. At each node on the shortest path from $R_3$ to $R_0$, $(s, g)$ state is created, the interface on which the $(s, g)$ join was received is added to the $(s, g)$ OIL, and an $(s, g)$ join is sent towards $R_0$.

4. When the last $(s, g)$ join reaches $R_0$ on interface $i_{11}$, this interface is added to the $(s, g)$ OIL at $R_0$. Since $R_0$ created $(s, g)$ state upon receiving the first packet from $s$, no state is created in this step.

5. Once the shortest path tree is established, $R_1$ sends an $(s, g)$ RP-bit Prune towards $RP$.

Typically, the same SPT-threshold is used on all routers, so each router with directly attached receiver hosts would initiate an $(s, g)$ join; the above steps detail this process only for $R_1$. Also, even though there is now a shortest path from $s$ to $R_1$, router $R_0$ continues to send $(s, g)$ packets to $RP$; this is required so that any new receiver node used by $g$ can learn, over the shared tree, about source host $s$.

**Procedure**: host $rcvr_1$ leaves group $g$:

1. $rcvr_1$ leaves $g$ by sending an IGMP leave group message to $R_1$.

2. $R_1$ receives the leave group message on interface $i_1$, and removes $i_1$ from its $(\star, g)$ OIL and $(s, g)$ OIL. Since the $(\star, g)$ OIL at $R_1$ is now empty, $R_1$ sends a $(\star, g)$ prune towards $RP$, out of interface $i_3$. Router $R_1$ also stops sending periodic $(s, g)$ joins to $R_3$.

3. $R_3$ receives the $(\star, g)$ prune on $i_5$, and drops $i_5$ from its $(\star, g)$ OIL and $(s, g)$ OIL; since the $(\star, g)$ OIL at $R_3$ is now empty ($rcvr_2$ already left), $R_3$ sends a $(\star, g)$ prune towards $RP$, out of interface $i_8$.

4. At each node on the shortest path from $R_3$ to $RP$, the incoming interface (on which the prune was received) is deleted from the $(\star, g)$ OIL, and a $(\star, g)$ prune is sent towards $RP$.

5. When the $(\star, g)$ prune finally reaches $RP$, on interface $i_9$, this interface is deleted from the $(\star, g)$ OIL at $RP$.

6. When an $(s, g)$ packet sent by $s$ on $SPT(s, g)$ is received by $R_3$ on interface $i_7$, node $R_3$ responds by deleting $(s, g)$ state and sending an $(s, g)$ prune towards $s$, out of interface $i_7$. The router upstream of $R_3$ (in the direction of $s$) responds by deleting $(s, g)$ state and sending an $(s, g)$ prune towards $s$. This process continues until finally $(s, g)$ state has been deleted on each node on the shortest path from $R_3$ to $R_0$.

### 3.7.3 BiDirectional PIM

Two drawbacks of PIM-SM are ($i$) the initial encapsulation of packets from a source host causes extra processing and delay, and ($ii$) the creation of $(s, g)$ state imposes extra memory and processing requirements beyond the

memory and processing needed for $(\star, g)$ state. These limitations are elimi-
nated in *BiDirectional PIM* (BiDir), proposed in 2007 by Handley, Kouvelas,
Speakman, and Vicisano [46]. BiDir is similar to CBT (Section 3.6), which
was proposed years earlier. For each $g$, BiDir creates a bidirectional tree,
rooted at some reachable address, which we again call the *RP*. Only $(\star, g)$
state (or sometimes no state at all) is imposed on each node in the bidirectional tree for $g$; no $(s, g)$ state is ever created. Thus if there are $G$ groups, a
maximum of $G$ $(\star, g)$ states are created at a node.

For BiDir, the RP might be the IP address of a router, or it might be the
address of a subnet. No physical router is needed to perform the RP functionality for BiDir. (In contrast, for PIM-SM, the RP must be an actual physical
router.) A different RP address may be selected for different groups; this
might be accomplished by partitioning the multicast address range among
different RPs. The RP can be configured statically, or advertised dynamically, using AutoRP or BSR (Section 3.8). The *RPF interface* with BiDir
has the same meaning as with PIM-SM; it is the interface towards the *RPF
neighbor*, which is the next node on the shortest path to the RP.

In order to prevent duplicate streams, on each subnet, and on each point-
to-point link, one *designated forwarder* (DF) router is elected for each RP.
The only exception to this rule is that if the RP itself is a subnet, no DF is
elected. The elected node serves as DF for all multicast groups mapping to
the RP. The router elected as DF is the one with the lowest unicast cost to
the RP, and ties are broken using the IP address of the router. The selection
of the DF is not a data driven event. Rather, for each RP, a DF is elected on
each subnet, and on each point-to-point link, before any packets arrive. (PIM-
SM uses a similar process, electing a *designated router* on each subnet.) For
a point-to-point link, the DF is simply the RPF neighbor. This is illustrated
in Figure 3.7 below, which shows a BiDir tree for some $g$. There are four

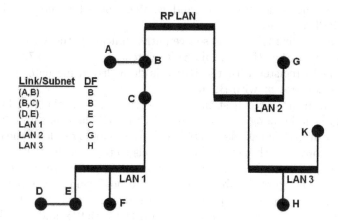

**Fig. 3.7** BiDir designated forwarder

subnets: the RP LAN, LAN 1, LAN 2, and LAN 3, and four point-to-point links: $(A, B)$, $(B, C)$, and $(D, E)$.

As with PIM-SM, in BiDir PIM join and prune messages are used to construct branches of the shared tree between the RP and receiver nodes. A node on a link or subnet wishing to receive $(\star, g)$ packets sends a join towards the DF for the link/subnet. In turn, this DF sends a join to the DF on its RPF interface, so on up the tree, until the RP, or some branch of the shared tree connecting to the RP, is reached. The join creates $(\star, g)$ state on each node encountered. Since packets always flow in the opposite direction of the join, this process creates a path from the RP to each receiver node. BiDir allows pruning of a leaf node when all subtending receiver hosts have left the group. As with PIM-SM, a prune for group $g$ travels hop-by-hop up the tree until it reaches the RP or some branch of the $(\star, g)$ shared tree connecting to the RP.

The only node on a link/subnet that can forward packets (arriving from an upstream node) onto that link/subnet is the DF for that link/subnet. A packet is accepted by a node $n$ on interface $i$ if $n$ is the DF for interface $i$; if the packet is accepted then $n$ $(i)$ sends the packet out of each interface in its $(\star, g)$ OIL, and $(ii)$ sends the packet upstream on the RPF interface towards the RP.

Step $(ii)$ above creates a path from each source node to the RP for $g$. No $(\star, g)$ state is created on a DF as a result of this forwarding; $(\star, g)$ state is only created on the DF as a result of receiving a PIM join (as described above) or an IGMP membership report (if a source host subtends the DF). It might be that for some branch of the $(\star, g)$ tree there are no receiver nodes for $g$. Such a branch is called a *source only* branch, and is used only for forwarding packets up towards the RP. Each node on a source only branch need only store the RPF interface for $g$, indicating the path to the RP for this $g$, and no $(\star, g)$ state is required. For example, in Figure 3.7, if there are no receiver hosts subtending nodes $C$, $D$, $E$, and $F$, then these four nodes need only store the RPF interface.

Thus with BiDir there are no source paths (unicast paths with $(s, g)$ state from a source to the RP), and no SPT-switchover (Section 3.7.2). Sources do not have to register with the RP; this data driven mechanism used by PIM-SM is not present with BiDir.

Packets from a source host flow up the tree towards the RP, but if at some node $n$ node along the path towards the RP there is a receiver host downstream of $n$, then packets will turnaround at $n$ and flow down the branch. However, a copy of each packet always travels to the RP, since there may be downstream receivers that can only be reached after first reaching the RP. For example, in Figure 3.1 above, suppose the bold solid lines represent the BiDir $(\star, g)$ tree. Consider a packet whose source node is node 9. To reach node 7, a packet will flow up to the RP (using the path 9-5) and then down to the receiver (using the path 5-2-3-7), even if the direct path 9-6-7 has lower cost. Now consider a packet whose source node is node 1. To reach node 3, it will not travel to the RP, but rather takes the path 1-2-3. However, node 2 will send a copy of the packet to the RP, where it is dropped if the $(\star, g)$ OIL at the RP is empty.

While BiDir reduces the amount of state relative to PIM-SM, since $(s, g)$ state is never created, it has some limitations. First, latency with BiDir may be higher than with PIM-SM, since there is no option to build shortest path trees from a source node to a receiver node. Second, BiDir does not allow more than one RP to be defined for each group (this is also true with PIM-SM). Third, inter-AS routing using MSDP (Section 5.2) will not work with BiDir, since MSDP advertises each $(s, g)$ to other domains, and BiDir does not maintain $(s, g)$ state. Fourth, anycast RP (Section 3.8.4), which allows multiple RPs to be defined, each advertising the same address, and where each node uses the closest RP, cannot be used with BiDir. However, BiDir does support RP redundancy, which sends all traffic to the secondary RP in case the primary RP fails or the RP LAN becomes unreachable.

## 3.7.4 Source Specific Multicast

PIM-SM and BiDir enable *any source multicast* (ASM), which means that once a receiver host signals (using an IGMP membership report) its desire to receive packets for some group $g$, it will receive packets originated by *any* source for $g$. The idea of PIM *Source Specific Multicast* (SSM), developed in 2006 by Holbrook and Cain [48] (see also [15]), is that a host may wish to receive packets from only a specific source host $s$ for a given group $g$, rather than receive packets from all source hosts for $g$. SSM builds a shortest path tree $SPT(s, g)$ rooted at the source node $n(s)$, the node that $s$ subtends. Relative to ASM, SSM reduces the load on the network and increases security.

SSM is well suited to one-to-many multicast applications such as network entertainment channels. A receiver host might learn about a specific $(s, g)$ by some out-of-band communication, e.g., by clicking on a browser link. Alternatively, the host might learn about $(s, g)$ if the host joined $g$ and the host subtends a receiver node which is running PIM-SM and received an $(s, g)$ stream.

With SSM, a host subtending node $x$ signals its desire to receive an $(s, g)$ stream by sending an IGMP membership report to $x$. Then an $(s, g)$ PIM join is sent node by node, starting at $x$, out of the RPF interface towards $s$, until we reach some node on $SPT(s, g)$ (possibly $n(s)$ itself). This creates a branch of the source tree spanning $x$. Since SSM builds a shortest path tree rooted at $n(s)$, no shared tree is built, and thus no RP is required. Since no shared tree is used, SSM creates only $(s, g)$ state, and never $(\star, g)$ state.

The ability for a receiver host to specify the desired $(s, g)$ is provided, for IPv4, by IGMP Version 3 (IETF RFC 3376) and, for IPv6, by *Multicast Lis-*

*tener Discovery* (MLD) Version 2. The address range for SSM is 232.0.0.0/8 for IPv4 addresses or FF3x::/32 for IPv6 addresses. To implement SSM, first PIM-SM must be configured, since SSM relies on the PIM-SM join/prune mechanisms. SSM can co-exist with PIM-SM, meaning both protocols can be running on a node, or SSM only can be deployed in a multicast domain.

## 3.8 Rendezvous Point Advertisement and Selection

With PIM-SM and BiDir, each PIM enabled node must learn the identity of the one or more RPs in the multicast domain. There are three popular methods used to accomplish this [112]: (*i*) Static RP, (*ii*) Auto-RP, and (*iii*) Bootstrap Router. We also discuss Anycast RP, used to select the closest RP in a PIM-SM domain. Since any router, or the network connections to it, may fail, typically multiple nodes are configured to be eligible to serve as an RP. We call such an eligible node a *candidate RP*.

### 3.8.1 Static RP

With *static RP*, the simplest method of informing each router about the RP identity, the RP address is manually configured in each node. The same RP can be configured for each node, or to balance the load, different RPs might be used for different sets of nodes. In the latter case, the set of RPs might be interconnected using MSDP (Section 5.2). One problem with this method is that, if the RP fails, the RP is now unreachable. This problem is overcome by selecting a second and possibly third backup RP, and configuring their addresses in each node. However, this manual configuration is onerous and error prone. Often, the best approach is to let routers automatically and dynamically determine the RP for each multicast group. The Auto-RP and Bootstrap Router methods are dynamic.

### 3.8.2 Auto RP

*Auto RP* [112] is a widely deployed but proprietary method, not an IETF standard. With Auto RP, each candidate RP announces itself to a specially configured designated node called a *mapping agent*. The mapping agent then distributes to each node in the network, using multicast, the set of candidate RPs. The Auto RP method itself uses two multicast groups.

A node that is a candidate RP for a particular group range sends this information to the *RP_announce* multicast group, which has the fixed group address 224.0.1.39. The mapping agent joins this group, and once it receives the candidate RP announcements, it selects one node to be the RP for each group range, thus populating the *Group-to-RP* table. The node chosen for each range is the node with the highest unicast address. If that node fails, the mapping agent selects, for each range, a new winner from among the remaining candidates. Once the mapping agent has made its selection, it sends the *Group-to-RP* table to the *RP_discovery* multicast group, which has the fixed address 224.0.1.40. Each node in the network joins this multicast group, and thus receives the *Group-to-RP* table. Backup mapping agents can be configured in case the active agent fails. If multiple mapping agents are configured, each agent operates independently, receiving the same set of candidate RP announcements, applying the same selection criterion, and hence advertising the same *Group-to-RP* table.

For a node $n$ to join the *RP_discovery* and *RP_announce* groups, it must know the addresses of the RPs for the shared tree for these two groups (it might be the same RP). The addresses are required so $n$ can send a join to the RPs. One way to do this is to configure these two groups to run PIM Dense Mode, so the *Group-to-RP* table sent by the mapping agent reaches every node. Rather than do this, the solution adopted is to configure a PIM option called *Sparse-Dense mode* on each node. This tells a node, for a given group $g$, to run PIM-SM if there is an entry for $g$ in the *group-to-RP* mapping table at the node, and to run PIM Dense Mode otherwise. This option allows a network administrator to initially use PIM Dense Mode for all multicast groups, and then, once the *Group-to-RP* table has been received by each node, easily switch to using PIM-SM for any desired group range.

### 3.8.3 Bootstrap Router

Whereas the Auto RP method utilizes multicast to the disseminate the *group-to-RP* mapping, the *Bootstrap Router* (BSR) method, part of the PIM Version 2 specification, does not utilize multicast to disseminate this mapping. With BSR, one node is selected as a *bootstrap router*. Each candidate RP sends its candidate RP announcement to the BSR using unicast routing. The set of such announcements received by the BSR is called the *candidate RP set*. The BSR then periodically floods the candidate RP set to all routers in the network, by sending it out of each interface, where it is forwarded, hop by hop, to all routers. Each router receiving the candidate RP set resends it out of all interfaces, except the incoming interface, with a Time to Live (TTL) value of 1 (so that each packet sent by a node expires when it reaches the adjacent node). The candidate RP set sent by the BSR contains the IP address of the BSR; this enables each candidate RP to know where to send its candidate RP announcements. Once a router has received a candidate RP set, for some configured period of time the router will discard any other candidate RP set received, unless the set was originated by a router with a higher IP address; if this occurs, the originating router is the new BSR, and the new candidate RP set is accepted and forwarded.

All routers in the network receive the same set of candidate RP announcements, and all routers execute the same hashing algorithm on this set to determine which candidate RP to select as the RP for each multicast group address range. The inputs to the hashing algorithm are a candidate RP address, a group address $g$, and a hash mask. Each router runs the hashing algorithm for each candidate RP that advertises its ability to serve as an RP for $g$. Since each router has the full set of announcements, if the current RP fails, each router can rerun the hashing method to determine a new RP.

BSR supports having multiple nodes serve as RPs, with each such node responsible for some range of multicast group addresses. However, only one node can be the designated RP for a given address. Should the current BSR fail, the BSR method provides several degrees of redundancy. First, multiple candidate BSRs can be configured, and the router with the highest BSR priority becomes the BSR. If the current BSR fails, the candidate with the next highest BSR priority takes over. If all candidate BSRs fail, then each router will use as its RP the statically configured RP, if one has been designated. Finally, if all candidate BSRs fail, and there is no statically configured RP, then each router will default to PIM Dense Mode multicast.

## 3.8.4 Anycast RP

The *anycast RP* technique allows multiple RPs to coexist and share RP duties within a single multicast domain running PIM-SM. To implement anycast RP, two or more RP nodes are configured with the *same* address. All nodes in the domain must learn this address, either through static configuration, or by using Auto-RP or BSR.

Now consider some source host $s$, subtending source node $n(s)$, wishing to send a multicast stream to some group. Recall that first $n(s)$ must register with the RP, by sending a unicast message. When multiple nodes are configured with the same address, the *Interior Gateway Protocol* (IGP) routing protocol (e.g., ISIS or OSPF) will route the message to the closest RP (i.e., the one whose IGP distance from $s$ is the smallest). Since each host sends to the closest RP, the RP load will in general be shared among the set of RPs. If one RP fails, then upon re-convergence of the IGP this RP will never be selected by any host, and its load will be taken up by the remaining RPs.

Recall that in a PIM-SM multicast domain, receiver nodes send a PIM join towards the RP, using the unicast routing table. If for a given group the source and receiver nodes are geographically distant, it is likely that they will connect to different RPs. To ensure that each source registering with one RP is known to the other RPs, the *Multicast Source Distribution Protocol* (MSDP) protocol is used. With MSDP, a full mesh of MSDP *peering sessions* between the RPs is created. If a source node $n(s)$ for some $(s, g)$ registers with one RP, a *source active* (SA) message is sent to all other RPs informing them of the existence of this $(s, g)$. We revisit MSDP in Section 5.2.

## 3.9 Comparison of PIM Methods

The major features of the four methods we have examined are summarized in Table 3.1, where "N/A" means "not applicable."

|  | PIM Dense Mode | PIM-SM | SSM | BiDir |
|---|---|---|---|---|
| **RP required** | no | yes | no | yes[1] |
| **shared tree** | no | yes | no | yes |
| **source tree** | yes | if SPT threshold exceeded | yes | no |
| **node state** | $(s, g)$ | $(s, g)$ on source path, $(\star, g)$ on shared tree $(s, g)$ on $SPT(s, g)$ | $(s, g)$ | $(\star, g)$, but no state on source only branch |
| **flood and prune** | yes | no | no | no |
| **supports anycast RP** | N/A | yes | N/A | no |

[1] An RP LAN address is sufficient, no physical node need be designated as RP

**Table 3.1** Comparison of PIM Protocols

## 3.10 Multipoint Trees using Label Switched Paths

There are many ways to dynamically create trees other than the PIM based methods studied above. Recently, considerable effort (e.g., [72]) has been devoted to building *Point-to-Multipoint* (P2MP) and *Multipoint-to-Multipoint* (MP2MP) trees using *Label Switched Paths* (LSPs). As these methods build upon the basic principles of *Multiprotocol Label Switching* (MPLS) ([43], [29], [30]), we begin by briefly reviewing this topic in the context of unicast routing of a packet.

The fundamental concept behind MPLS is that each node maintains a forwarding table of the following form: if node $n$ receives a packet with label $L$, then $n$ swaps (i.e., changes) the label to $L'$ and sends the packet out the arc associated with $L$. The labels $L$ and $L'$ are chosen so that the packet takes the desired path through the network. To see how this can be applied to shortest path routing, suppose that a shortest path between nodes $s$ and $d$ visits the sequence of nodes $n_0, n_1, n_2, \cdots, n_k, n_{k+1}$, where $n_0 = s$ and $n_{k+1} = d$. For $j = 0, 1, 2, \cdots, k$, let $a_j$ be the arc from $n_j$ to $n_{j+1}$ on the shortest path. Let $L_0, L_1, \cdots, L_k$ be any $k$ labels; they may be chosen arbitrarily. For $j = 0, 1, 2, \cdots, k$, we associate label $L_j$ with arc $a_j$ on node $n_j$. For $j = 0, 1, 2, \cdots, k$, let node $n_j$ store the following *label swapping rule*: if a packet arrives with label $L_j$, then change its label to $L_{j+1}$ and send it out on arc $a_j$. Now consider a packet arriving at node $s$ (e.g., from a subtending source host) with destination $d$ and no label. Node $s$ examines it *Forwarding Information Base* (FIB) and sees that a packet with destination $d$ should receive the label $L_1$. So $s$ pushes (i.e., imposes) the label $L_1$ on the packet and sends it out on arc $a_0$. The packet is received by $n_1$, which examines its LFIB, which says that any packet with label $L_1$ should have its label swapped with $L_2$ and be sent out on arc on $a_1$. So $n_1$, following its LFIB, changes the label to $L_2$ and sends the packet out on arc $a_1$. The packet is received by $n_2$, which changes the label to $L_3$ and sends the packet out on arc $a_2$. Ultimately, the packet is received by $n_k$, which changes the label to $L_{k+1}$ and sends the packet out arc on $a_k$. The packet now arrives at the destination $d$.

Note that it is not necessary for $n_k$, the predecessor node to $d$, to impose on a packet the final label $L_{k+1}$ for the final hop from $n_k$ to $d$. Rather, since each node can examine the packet to determine the destination $d$, node $n_k$ can determine that $d$ is the next hop. Therefore $n_k$ can simply forward the packet out on arc $a_k$ towards $d$, without imposing another label. When $d$ receives the packet, $d$ determines that it is the destination node, and $d$ forwards the packet for final processing (e.g., to the subtending host). The action taken by the penultimate node $n_k$ (one hop away from the destination) of removing a label, rather than label swapping, is called *penultimate hop popping*. There are many applications of MPLS in which multiple layers of labels are imposed on a packet. In the shortest path example above, the label stack is only one layer deep.

A *label switched router* (LSR) is a router with the ability to impose (push) a label on a packet, swap one label for another, or remove (pop) a label from a packet. In the above shortest path example, node $s$ pushed a label, nodes $n_1$ to $n_{k-1}$ swapped labels, and $n_k$ (with penultimate hop popping) popped a label. When MPLS is used in conjunction with a routing protocol such as OSPF, the label swapping rule will cause each packet to take the shortest path to its destination. One big advantage of using MPLS rather than pure OSPF routing is that MPLS can be used to force a packet to take a specific path, e.g., along longer high capacity arcs, and MPLS facilitates fast reroute techniques, to deal with an arc or node failure. In the above example of a shortest path from $s$ to $d$, the source node $s$ pushed the label $L_1$ for packets destined for $d$. In general, an LSR $n$ will store, in its LFIB, a label to impose on a packet for each IGP next hop known to $n$. It will also store in its LFIB the label swapping $(L, L')$ map to apply to any packet received with incoming label $L$; it swaps $L$ for $L'$ unless $n$ is the penultimate node. A label has local significance only, so that two different LSRs are free to advertise the same label.

We are now ready to apply the concepts of MPLS to multicast routing. An *egress* node is an LSR that can remove a packet from the LSP for native IP processing, e.g., to send to a receiver host. A *transit* node is an LSR which has a directly connected upstream (towards the root of a tree) LSR and which also has directly connected downstream LSRs. A P2MP LSP contains a root node, one or more *egress* nodes, and a set (possibly empty) of transit nodes. A P2MP LSP is uniquely defined by the ordered pair $(\bar{n}, v)$, where $\bar{n}$ is the address of the root node, and $v$ is an integer *opaque value*, such that no two P2MP LSPs rooted at $\bar{n}$ have the same opaque value. A node $y$ on the LSP $(\bar{n}, v)$ maintains a *label map list*, which has the form $L \rightarrow \{(a_1, L_1), (a_2, L_2), \cdots, (a_k, L_k)\}$. This says that if $y$ receives a packet with label $L$, then it makes $k$ copies. For the first copy, label $L$ is swapped with $L_1$ and the packet is sent out on arc $a_1$; for the second copy, $L$ is swapped with $L_2$ and the packet is sent out on arc $a_2$, etc. Suppose the root $\bar{n}$ has the label map list $NULL \rightarrow \{(a_1, L_1), (a_2, L_2), \cdots, (a_k, L_k)\}$ for $(\bar{n}, v)$. This means that, if an unlabelled packet arrives at the root $\bar{n}$, then $\bar{n}$ makes $k$ copies, sending the first copy out on $a_1$ with label $L_1$, the second out on $a_2$ with label $L_2$, etc.

Now suppose a node $x$ wants to join the P2MP tree $(\bar{n}, v)$. It first determines the next hop node $y$ on the best path (as determined by unicast routing) from $x$ to $\bar{n}$. Node $y$ is called the *upstream LSR* for $x$ for $(\bar{n}, v)$. Next, $x$ allocates a label $L_x$, and sends a *label map message* $(\bar{n}, v, L_x)$ to $y$ over the arc $a_x$ connecting $x$ to $y$. Assume that $x$ is not the upstream LSR of $y$ for $(\bar{n}, v)$. Node $y$ checks whether it already has $(\bar{n}, v)$ state. If it does, then $y$ updates its label map list to include $(a_x, L_x)$, so the new label map list at $y$ for $(\bar{n}, v)$ is $L \rightarrow \{(a_1, L_1), (a_2, L_2), \cdots, (a_k, L_k), (a_x, L_x)\}$. If $y$ does not have $(\bar{n}, v)$ state, then $(i)$ $y$ allocates a label $L$, $(ii)$ $y$ installs the label map list $L \rightarrow \{(a_x, L_x)\}$ for $(\bar{n}, v)$, and $(iii)$ $y$ sends the label map message $(\bar{n}, v, L)$ to its upstream LSR for $(\bar{n}, v)$. Step $(iii)$ is required since $y$ is not yet on the P2MP LSP $(\bar{n}, v)$.

When a label map message $(\bar{n}, v, L)$ reaches $\bar{n}$ from node $x$ over arc $a_x$, the root determines if it has $(\bar{n}, v)$ state. If it does, then $\bar{n}$ updates its label map list for $(\bar{n}, v)$ to $NULL \rightarrow \{(a_1, L_1), (a_2, L_2), \cdots, (a_k, L_k), (a_x, L)\}$, so an extra copy is sent out on $a_x$ to node $x$ with label $L$. If not, then $\bar{n}$ creates the label map list $NULL \rightarrow \{(a_x, L)\}$ for $(\bar{n}, v)$.

In the creation of the P2MP tree, each leaf node originates the label that is sent up the tree towards the root; this is called *downstream label assignment*. These label map messages and procedures are very similar in functionality to PIM join messages and procedures. Similarly, [72] defines label withdraw messages and procedures analogous to PIM prunes.

So far we have shown how to build a P2MP tree rooted at a given node $\bar{n}$, where $\bar{n}$ is any LSR, and $\bar{n}$ need not be a source or receiver node for $g$. This approach can be used to create a shared tree, for a given multicast group $g$, to support any-to-any multicast. The tree spans the source and receiver nodes for $g$. Packets flow from source nodes to the root via unicast routing, e.g., using a multipoint-to-point (MP2P) (as opposed to P2MP) LSP rooted at $\bar{n}$. The MP2P tree provides an LSP from each node used by $g$ to $\bar{n}$. When the packets reach $\bar{n}$, they are sent down the P2MP tree rooted at $\bar{n}$. For the MP2P LSP, the label assignment starts at the root node $\bar{n}$, not at the leaf nodes; this is called *upstream label assignment*. This approach to building a shared tree will result in each source node receiving, over the P2MP tree, any packet it sent towards $\bar{n}$, so a source node must recognize such duplicate packets and discard them. This approach can be viewed as a generalization of the PIM-SM approach of creating a shared tree rooted at the RP over which packets flow downstream, together with a source path from each source node to the RP.

LSP techniques can also be used to create an MP2MP tree for $g$. An MP2MP tree also has a single root node $\bar{n}$. Define a *leaf node* of an LSP to be a node that either imposes a label on a packet or removes a packet from an LSP for further processing (e.g., to send to a directly attached host). To build an MP2MP tree, each leaf node establishes both a downstream P2MP and an upstream LSP which is used to send packets towards the root and other leaf nodes. Packets from a downstream (relative to the root $\bar{n}$) node follow the upstream LSP towards the root, but may also follow a downstream branch (to reach receiver nodes) before reaching $\bar{n}$. A packet from $\bar{n}$ destined to a receiver node is forwarded just as with the P2MP tree rooted at $\bar{n}$. The details are provided in [72]. Finally, we note that the current specification for using the Label Distribution Protocol (LDP) to set up P2MP and MP2MP trees assumes that LDP neighbors are directly connected; Napierala, Rosen, and Wijnands [79] show how *targeted LDP* can be used to create P2MP and MP2MP trees when the LDP neighbors are not directly connected.

# Chapter 4
# Other Multicast Routing Methods

In this chapter we review some other approaches to multicast. We contrast application versus overlay methods, describe a gossip based system motivated by peer-to-peer file sharing techniques, and present a method based upon distributed hashing. We describe a method enabling hosts with only unicast capabilities to access multicast content from the Internet. Lastly, we present an algorithm for constructing a tree subject to a delay constraint, and a routing method for wireless networks inspired by the foraging of ants.

## 4.1 Application and Overlay Multicast

Thus far we have assumed that source hosts and receiver hosts are directly connected to routers, and it is the routers which have the responsibility for creating the multicast tree and for forwarding packets along the tree. Calling this model the *IP multicast* approach, Lao, Cui, Gerla, and Maggiorini [65] note that IP multicast is not the only possible approach to Internet multicast. In *application level* multicast, the hosts themselves are responsible for tree creation and packet forwarding. They further classify application level multicast methods as either structured methods, which use distributed hash tables and build trees on top of a structured overlay, or unstructured methods, which either organize hosts into a mesh and build trees on top of this mesh, or directly construct a tree or some other topology. In *overlay multicast*, additional nodes, called proxy or services nodes, are strategically deployed in the network, and a backbone overlay network connects the proxy nodes. Each host communicates with its closest proxy. This overlay network can be a tree, to provide efficient multicast routing, and the limited size of the overlay network reduces network management overhead at proxy nodes.

These three approaches are illustrated in Figure 4.1, for the same set of 7 hosts. In the IP multicast network (middle figure), there are 5 routers. In the overlay network (top figure), 4 proxy nodes are deployed; each host connects to the closest proxy, and an overlay network, with 3 links, connects the proxies. These 3 links do not necessarily correspond to actual physical links, and instead might represent tunnels between routers in the IP multicast network. In the application layer network (bottom figure), 6 links interconnect the hosts; these links also might be tunnels between nodes in the IP multicast network.

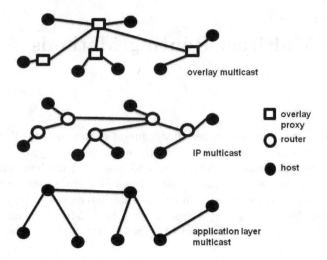

**Fig. 4.1** Approaches to overlay multicast

Some of the drawbacks of application level multicast are: (*i*) Hosts are less reliable than routers, so the trees might be more unstable. (*ii*) Since hosts do not know the full network topology, the trees created are typically longer than with IP multicast. (*iii*) A separate tree must be created for each group or application, which is not the case with IP multicast or overlay networks.

Simulations were used to compare the approaches, both at the router level (where an arc is a connection between routers), and at the AS level (where nodes represents ASs, and an arc represents a connection between two routers in different ASs). Group members were geographically distributed in the network using a uniform random distribution. SSM (Section 3.7.4) was used to generate the IP multicast trees, so the trees are shortest path trees. For the overlay network, nodes with the highest degree were selected as proxies (the simulations studied the impact of the number of proxies). The results show that IP multicast trees have the lowest cost, application layer trees have the highest cost, and overlay trees are inbetween. IP multicast has the same delay as unicast (since shortest path trees are used), overlay has slightly higher delay than IP multicast, and application layer has the highest delay. IP multicast has the least control overhead; application layer has smaller control overhead than overlay for small groups or large networks, but its overhead exceeds that of overlay when the group size gets sufficiently large. They conclude that overlay is a promising approach, especially for an ISP, while application layer is good for quick deployment, especially for small groups.

We now consider one example of an unstructured application layer protocol, a gossip based method, and one example of a structured application layer protocol, employing content addressable networks.

## 4.2 A Gossip Based Method

Chainsaw [82] is a multicast protocol motivated by gossip based protocols and the peer-to-peer file sharing system *BitTorrent*. As packets are generated by a given source, they are given a series of increasing sequence numbers. When a source generates packets, it sends a *NOTIFY* message to its neighbors, and when a node receives a packet, it sends a *NOTIFY* message to all its neighbors. Each node maintains a sliding *window of interest*, which is the range of sequence numbers of packets that the node is interested in receiving at the current time. Each node also maintains a sliding *window of availability*, which is the range of sequence numbers of packets that the node is willing to forward to a neighbor. Typically, the window of availability is larger than the window of interest. A node $n$ maintains, for each neighbor, a list of packets that the neighbor has, and also a list of desired packets that $n$ wants to obtain from the neighbor and that are within the neighbor's window of availability. Node $n$ then applies some criteria to pick one or more packets from this list, and requests them using a *REQUEST* message. (A simple strategy is to pick randomly from the list.) Node $n$ keeps track of which packets it has requested from each neighbor, ensures that the same packet is not requested from multiple neighbors, and limits the number of unfulfilled requests made to each neighbor. Each node also stores a list of requests from its neighbors, and sends the requested packets if there is sufficient bandwidth. The algorithms that nodes use to update their window of interest and window of availability, and to determine when to forward packets, are application specific. For example, if the application requires strict ordering of packets, packets would be forwarded only after a contiguous block of sequence numbers has been received. It might happen that packets are not requested from a source until several seconds after they were generated. To ensure that such packets are forwarded to neighbors, and not lost, the source can maintain a list of packets that have not yet been forwarded. If the list is not empty and the source receives a request for a packet not on this list, the source ignores the sequence number requested and instead sends the oldest packet on the list, and deletes that sequence number from the list.

## 4.3 Multicasting using Content Addressable Networks

A novel approach to multicast, developed in 2001 by Ratnasamy, Handley, Karp, and Shenker [88], utilizes the method of Ratnasamy, Francis, Handley, Karp, and Shenker [87] for routing in a *Content Addressable Network* (CAN). CANs were developed to scale peer-to-peer file sharing systems (e.g., to download music), for which it was recognized that the performance bottle-

neck is determining which peer is storing the desired file. The CAN solution
is to extend hashing techniques, and let nodes in coordinate space provide
the hashing functionality of inserting, lookup, and deletion of $(key, value)$
pairs. The hash table is partitioned among the nodes; the chunks of the partition are called *zones*. Each node stores a zone, and a zone is associated with
exactly one node. Each node also knows the identity of a small number of
adjacent zones. A request for an insertion/lookup/deletion corresponding to
a given *key* is routed through the CAN coordinate space until the request
reaches the zone containing *key*. In the context of file sharing, the *key* is the
file name, and the *value* is the IP address of the node storing the file.

The CAN coordinate space is a $d$-torus, rather than simply $d$-dimensional
Euclidean space, so that the coordinates wrap around. Let $\mathcal{X}_d$ denote the
$d$-torus. (For example, if $d = 2$, a 2-torus has the familiar shape of a donut,
obtained by rotating a circle around an axis that does not intersect the circle.)
To add the pair $(key, value)$ to a CAN, first *key* is mapped to a point $P \in \mathcal{X}_d$
using a deterministic hash function. Let $x$ be the node that owns the zone in
which $P$ lies. Then $(key, value)$ is added to the memory at $x$.

Suppose some node $n$ wants to retrieve the *value* corresponding to a given
*key*. The same hashing function is applied to *key* to map *key* to some point
$P \in \mathcal{X}_d$. If $P$ is in the zone owned by $n$, then $n$ can immediately retrieve
*value*, since $n$ knows $(key, value)$ for points in its zone. If $P$ is not in the
zone owned by $n$, then $n$ must route the request through the CAN until the
request reaches the node who owns the zone in which $P$ lies.

In $d$-dimensional space, two zones are neighbors if their coordinate ranges
overlap along $d - 1$ dimensions, and abut (i.e., line up) in 1 dimension. Referring to Figure 4.2, zones 1 and 4 are neighbors, since they overlap along the $x$
axis and abut along the $y$ axis. The neighbors of zone 1 are zones 2, 4, 7, 8, 9,

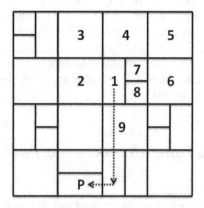

**Fig. 4.2** Content Addressable Network

and the neighbors of zone 7 are 1, 4, 6, 8. Unicast routing from a source $s$ to
destination $d$ is done by greedy forwarding to the neighbor with coordinates

closest to the destination coordinates, as illustrated by the dotted lines in Figure 4.2. For a $d$-dimensional routing space partitioned into $m$ equal sized zones, each node has $2d$ neighbors, and the average path length (in coordinate space) is $(d/4)(m^{1/d})$ hops [88]. However, each hop in coordinate space will in general correspond to many hops in the underlying IP network; some techniques for reducing the actual latency in CAN routing are given in [88].

When a new node $n$ joins the CAN, it must be given its own zone. To accomplish this, $n$ randomly chooses a point $P \in \mathcal{X}_d$ (using the *bootstrap* procedure described below) and sends, to any existing CAN node, a *join* request destined for $P$. This message is routed through the CAN until it reaches the zone $Z(P)$ in which $P$ lies. The current owner, say $n(P)$, of zone $Z(P)$ splits this zone in half and assigns half to node $n$. For example, a zone might first be split along the $x$ dimension, with the second split along the $y$ dimension, etc. The stored $(key, value)$ values from the half zone currently belonging to $n(P)$ are now transferred to $n$. The new node $n$ creates a list of its neighbor zones, node $n(P)$ updates its list of neighbor zones, and the neighbors of the zone $Z(P)$ that was split must be notified so that they can update their list of neighbors. When a node $n$ leaves the CAN, one of its neighbors is chosen to now own the zone that $n$ had owned, and the $(key, value)$ entries stored at $n$ are given to that neighbor to store.

The *bootstrap* procedure assumes that the CAN has a *Domain Name Server* (DNS) domain name, which resolves to the IP address of one or more CAN bootstrap nodes. Each bootstrap node stores a list of a few nodes that are presumed to be currently in the CAN. A new node $n$, to join the CAN, retrieves the bootstrap node's IP address from its DNS name. The bootstrap node supplies to $n$ several randomly chosen nodes in the CAN, and one of these is randomly selected by $n$ to receive its join message.

To now apply these concepts to multicast, suppose all the nodes in the network belong to a given CAN. We form a sub-CAN $C_g$ whose nodes are the source and receiver nodes used by group $g$. To create $C_g$, a hash function maps $g$ to a point $P_g \in \mathcal{X}_d$. The node $n_g$ in the CAN owning the zone containing $P_g$ becomes the bootstrap node for $C_g$. Each source or receiver node used by $g$ must join $C_g$, using the bootstrap node $n_g$. Multiple bootstrap nodes can also be utilized, by using multiple hash functions to deterministically map $g$ to multiple points in $\mathcal{X}_d$.

Within $C_g$, multicast forwarding is accomplished by flooding. Assume $C_g$ is a $d$-dimensional CAN, so each node has at least two neighbors (one in the forward direction and one in the reverse direction) along each of the $d$ dimensions. The flooding rules are as follows. ($i$) The source node forwards a packet to all its neighboring zones. (In Figure 4.3, the source $P$ sends to all neighbors.) ($ii$) If node $x$ receives a packet from node $y$, and $x$ and $y$ abut along dimension $i$, then $x$ forwards the packet to those neighbors which it abuts in dimensions $1, 2, \cdots, i-1$, and also to the neighbor which it abuts in dimension $i$, but in the opposite direction from which it received the packet. (In Figure 4.3, the node owning zone 2 receives the packet along

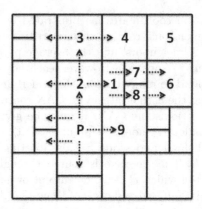

**Fig. 4.3** Flooding in a CAN

dimension 2, so it sends it out both directions in dimension 1 and in the opposite direction from $P$. And the node owning zone 1 receives the packet along dimension 1, so it sends it out only in dimension 1, in the opposite direction from zone 2.) (*iii*) To prevent a packet from looping around $\mathcal{X}_d$, a node does not forward a packet along a particular dimension if the packet has already gone at least halfway across $\mathcal{X}_d$, as determined from the source coordinate (in that dimension) of the packet. (*iv*) A node caches the sequence numbers of packets it has received and does not forward a packet that it has previously received.

The above rules ensure that, if all the CAN zones are equal size, each node will receive a node exactly once. However, with zones of different sizes, a node may receive multiple copies of a packet (e.g., zone 6 will receive the same packet from zones 7 and 8). An additional refinement to these flooding rules that reduces duplicate packets is presented in [88].

The advantage of using a CAN for multicast distribution is that the storage required at each node is small, routing is simple, and thus the CAN can scale to thousands of nodes. Increasing the dimension $d$ of the CAN coordinate space reduces the routing path length (in the CAN coordinate space) at the expense of a small increase in the number of neighbors each node must store. However, since the CAN is unaware of the underlying IP topology, the flooding may still utilize more bandwidth than tree-based IP methods; numerical results are provided in [88].

## 4.4 Automatic IP Multicast without Explicit Tunnels

Referring to Figure 4.4, consider the following Internet scenario. A content provider (e.g., streaming live sports coverage), labelled $C$ in the figure, uses a multicast enabled transit ISP (labelled $T$) to send the stream to two local ISPs: one, labelled $M$, is multicast enabled, and the other, labelled $U$, supports only unicast. In domain $U$, receiver hosts $r_1$ and $r_2$ subtend node $p$, and

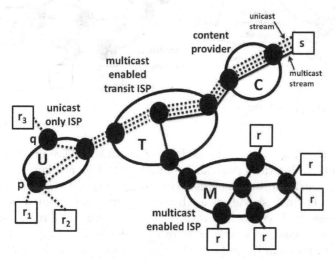

**Fig. 4.4** Without AMT: unicast connections to receivers in $U$

receiver host $r_3$ subtends node $q$. Since domains $T$ and $M$ support multicast, a multicast protocol (e.g., SSM) can be used to build a tree (illustrated by the solid lines) to deliver the stream to receiver hosts in $M$. However, since domain $U$ does not support multicast, it has no way (e.g., using a PIM join) to join a multicast tree in $T$. Rather, it is necessary to provide an end-to-end unicast connection (illustrated by the dotted lines) from the source $s$ to each receiver host in $U$. For high bandwidth streams with many receiver hosts in $U$, this is expensive (see Section 1.3). *Automatic IP Multicast without Explicit Tunnels* (AMT), developed in 2001 by Thaler, Talwar, Vicisano, and Ooms [105], provides an alternative approach which allows multicast to reach receivers in $U$ without explicit unicast tunnels, and without additional overhead to the routers in $T$. Stated simply, AMT allows anyone on the Internet to create a dynamic tunnel to download multicast data streams ([53], [60], [62]).

We describe AMT using Figure 4.5, which for simplicity considers only the receiver host $r_1$ in $U$. To implement AMT, domain $T$ configures one or more multicast enabled routers to serve as an *AMT relay*. In Figure 4.5,

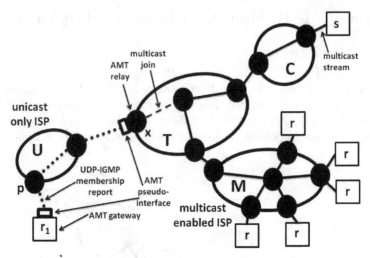

**Fig. 4.5** AMT procedures

node $x$ is the AMT relay. The address of the AMT relay must be known throughout the Internet; this can be accomplished by configuring the relay with an *anycast* address which is advertised throughout the Internet. Any packet with an anycast address as the destination will be routed by the IGP to the nearest node configured with this address. An AMT relay has one or more multicast enabled interfaces to other nodes in $T$, zero or more unicast interfaces to nodes in $U$, and an *AMT pseudo-interface*, which is the logical point at which multicast packets are encapsulated inside unicast packets.

An *AMT gateway* is a host or router, without native multicast connectivity to $T$, that sends AMT requests to the AMT relay. The AMT request leaves the AMT gateway on an AMT pseudo-interface. In Figure 4.5, the AMT gateway is configured on the host $r_1$.

When $r_1$ wants to receive an $(s,g)$ stream for some $g$, it originates an IGMP membership report destined for the first hop router $p$. The AMT gateway running on $r_1$ intercepts this report and initiates an anycast AMT request, towards the AMT relay $x$, to set up an AMT tunnel. Once the AMT tunnel is established, $r_1$ encapsulates, in *User Datagram Protocol* (UDP), the IGMP membership report into the AMT tunnel. The encapsulated report is received by the AMT relay, which decapsulates it, and issues a PIM join towards $s$. Thus, once the tunnel is established, the connection between the host $r_1$ and the AMT relay $x$ is equivalent to a standard IGMP Version 3 connection between the host and the AMT relay. The AMT gateway sends periodic *AMT Membership Update* messages to refresh the state on the AMT relay, and sends the appropriate message to leave $g$ when the $(s,g)$ stream is no longer desired.

This discussion considered only $r_1$; hosts $r_2$ and $r_3$ in $U$ would each create, in the same manner, its own AMT pseudo-interface to the AMT relay.

## 4.5 Distributed Delay Constrained Multicast Routing

Sometimes, e.g., for audio or video streams, the total delay from a source node $s$ to a receiver node $t$ cannot exceed a specified bound $\Delta$. Suppose the problem is to compute a least cost path from $s$ to $t$ subject to this delay constraint. A path is *feasible* if the total delay on the path does not exceed the bound. For such constrained routes, the next hop from some node $n$ will in general depend on the set of arcs traversed from $s$ to $n$. This is illustrated in Figure 4.6. Associated with each arc is the pair $(c, d)$, where $c$ is the cost

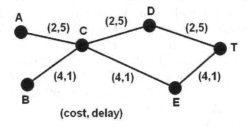

**Fig. 4.6** Delay constrained routing

and $d$ is the delay. Suppose $\Delta = 12$, and consider a packet arriving at $C$ destined for $T$. If the source is $A$, then the minimum cost path $A - C - D - T$ is not feasible, since its delay is 15, so $C$ must send the packet to $E$; the cost of path $A - C - E - T$ is 10 and the delay is 7. If the source is $B$, then $C$ should send the packet to $D$; the cost of path $B - C - D - T$ is 8, and the delay is 11. Thus, in this example of constraint based routing, the next hop at $C$ depends on the path taken by the packet prior to reaching $C$. For shortest path routing, in the absence of any constraints the next hop from a given node depends only on the destination of the packet, and not on the path taken by the packet prior to reaching the node.

Delay is just one *quality of service* (QoS) constraint. QoS constraints can be *additive* (summed over all links in a path), e.g., delay, cost, and jitter, or *min/max* (where the *min/max* is taken over all links in a path), e.g., bandwidth. For *min/max* constraints, any link not providing the required resource can simply be pruned from the network, so we assume this has been done, and consider only additive constraints. Both centralized and distributed heuristic methods have been proposed to solve, or approximately solve, the problem of computing a minimum cost path subject to one or more additive constraints; many are described by Ben Ali, Belghith, Moulierac, and Molnar [13]. One approach to this problem is to compute, for each receiver node, a constrained shortest path from the source node to the receiver node. The union of these paths forms a subgraph which in general must be pruned to eliminate cycles without violating the constraints. Greedy and taboo search procedures for pruning have been proposed [13].

Jia [54] proposed in 1998 a tree based method for multicast routing from a single source node to multiple receiver nodes, subject to a single additive constraint. The given data is an undirected graph $(\mathcal{N}, \mathcal{A})$, the cost $c_{ij}$ and delay $d_{ij}$ of arc $(i, j) \in \mathcal{A}$, a source node $s \in \mathcal{N}$, and a set of receiver nodes $\mathcal{D} \subseteq \mathcal{N}$. For $t \in \mathcal{D}$, let $P(s, t)$ be the unique path from $s$ to $t$ in a tree. Then the *delay constraint* is $\sum_{(i,j) \in P(s,t)} d_{ij} \leq \Delta$ for $t \in \mathcal{D}$. The problem is to construct a tree $\mathcal{T}$, rooted at $s$ and connecting to each $t \in \mathcal{D}$, such that the tree cost is minimized subject to the delay constraint. For brevity, by a *shortest* path we mean a *minimal cost* path.

Jia's method assumes that each node stores the minimal cost to each other node (and a corresponding shortest path), and also the delay of the shortest path computed. Two assumptions are required. The first assumption is that, for any two arcs $a_1$ and $a_2$, if $c(a_1) \geq c(a_2)$ then $d(a_1) \geq d(a_2)$. Under this strong assumption, the shortest path between any two nodes is also the least delay path between the two nodes. For $t \in \mathcal{D}$, let $\mathcal{P}_d^\star(s, t)$ be minimal delay path from $s$ to $t$. The second assumption is that for $t \in \mathcal{D}$ we have $\sum_{a \in \mathcal{P}_d^\star(s,t)} d_a \leq \Delta$. This assumption guarantees the existence of a tree satisfying the delay constraint.

In Jia's distributed method, each node operates independently, executes exactly the same algorithm, using only local information, and each node exchanges routing updates with neighboring nodes. The method starts by initializing $\mathcal{T} = \{s\}$ (i.e., initially $\mathcal{T}$ contains just the source node, and no arcs) and then growing $\mathcal{T}$. If node $t$ is not in $\mathcal{T}$, by a shortest path from $t$ to $\mathcal{T}$ we mean a path $P(t, n)$, using arcs in $\mathcal{A}$, such that $n \in \mathcal{T}$ and the cost of this path does not exceed the cost of any other path from $t$ to a node in $\mathcal{T}$. If $P(t, n)$ is a shortest path from $t$ to $\mathcal{T}$, we call $n$ the *tree node* closest to $t$.

The method can be summarized as follows. Starting with $\mathcal{T} = \{s\}$, we pick the node $t_1 \in \mathcal{D}$ that is closest to $s$; the shortest path from $t_1$ to $s$ is added to $\mathcal{T}$. We then pick the node $t_2 \in \mathcal{D}$ such that $t_2$ is not in $\mathcal{T}$ and $t_2$ is closest to $\mathcal{T}$ under the delay constraint; we add to $\mathcal{T}$ the shortest path from $t_2$ to the tree node closest to $t_2$. This process continues, where at each iteration we pick the node $t$ not in $\mathcal{T}$ which is closest to $\mathcal{T}$ under the delay constraint, and then add to $\mathcal{T}$ the shortest path from $t$ to the tree node closest to $t$. This continues until all nodes in $\mathcal{D}$ have been added to $\mathcal{T}$. It is proved in [54] that this method generates a tree spanning the nodes of $\mathcal{D}$ and satisfying the delay constraint.

The method uses three data structures. The structure $T2D$ (tree to destination), with global scope, has 3 fields: for $t \in \mathcal{D}$, the field $T2D[t].cost$ is the cost from $\mathcal{T}$ to $t$, the field $T2D[t].treenode$ is the tree node closest to $t$, and $T2D[t].inTree$ is 1 if $t \in \mathcal{T}$ and 0 otherwise. Second, for $t \in \mathcal{T}$, node $t$ stores $delay[t]$, the total accumulated delay in the path in $\mathcal{T}$ from $s$ to $t$. Finally, each node $n$ maintains a routing table $RT_n$, where if $u \in \mathcal{N}$, then $RT_n[u].cost$

is the cost of the shortest (unicast) path from $n$ to $u$, and $RT_n[u].delay$ is the delay of that shortest path to $u$.

By $\mathcal{D} - \{s\}$ we mean the set $\{t \mid t \in \mathcal{D}, t \neq s\}$. The source node $s$ does the following:

(*i*) For $t \in \mathcal{D} - \{s\}$, initialize $T2D[t].treenode = s$ and $T2D[t].inTree = 0$ (since $s$ is the only node in $\mathcal{T}$), and initialize $T2D[t].cost$ to the cost of the shortest path from $s$ to $t$.

(*ii*) Let $t_1 \in \mathcal{D}$ be any node such that $t_1$ is closest to $\mathcal{T}$. Add $t_1$, and all the arcs on the shortest path from $s$ to $t_1$, to $\mathcal{T}$.

(*iii*) Let $v$ be the successor node to $s$ on the shortest path from $s$ to $t_1$. Set $delay[v] = RT_s[v].delay$ (the delay from $s$ to $v$ on this shortest path).

(*iv*) Send a *setup* message from $s$ to $v$ with the parameters $delay[v]$, $T2D$, and $\mathcal{D}$.

Let $\mathcal{D} - \mathcal{T}$ be the set of nodes in $\mathcal{D}$ but not in the tree $\mathcal{T}$. Now consider a node $v$ receiving a *setup* message. Upon receiving the message, $v$ does the following:

(*v*) Determine, for each $t \in \mathcal{D} - \mathcal{T}$, if (*a*) $RT_v[t].cost < T2D[t].cost$ (i.e., if the cost from $v$ to $t$ is less than the cost from the current $\mathcal{T}$ to $t$), and also if (*b*) $delay[v] + RT_v[t].delay \leq \Delta$ (i.e., if the delay from $s$ to $v$, plus the delay from $v$ to $t$, is less than the bound $\Delta$).

(*vi*) If both (*a*) and (*b*) are true for some $t \in \mathcal{D} - \mathcal{T}$, then the path from $s$ to this $t$ through $v$ is a feasible and shorter path than the current path, so $v$ makes the updates $T2D[t].cost = RT_v[t].cost$ and $T2D[t].treenode = v$ (the closest node in $\mathcal{T}$ to $t$ is $v$).

(*vii*) Node $v$ sends a *setup* message to the next node, say $w$, on the shortest path from $s$ to $t_1$. The parameters of the *setup* message are $delay[w] = delay[v] + RT_v[w].delay$ (the delay from $s$ to $v$ plus the delay from $v$ to $w$), $T2D$, and $\mathcal{D}$.

Steps (*v*)-(*vii*) continue until $t_1$ has been reached and has executed steps (*v*)-(*vii*). At this point:

(*viii*) Pick as the next receiver node $t_2$ the node in $\mathcal{D} - \mathcal{T}$ which is closest to $\mathcal{T}$, i.e., the node $t$ in $\mathcal{D} - \mathcal{T}$ for which $T2D[t].cost$ is smallest. Let $z = T2D[t_2].treenode$ (the node in $\mathcal{T}$ closest to $t_2$). Note that, in general, $z$ is a *fork* node, meaning that there are more than two arcs in $\mathcal{T}$ incident to $z$. Add to $\mathcal{T}$ the path from $t_2$ to $z$.

(*ix*) Send a *fork* message to $z$, instructing $z$ to send a *setup* message to the next node on the shortest path from $z$ to $t_2$, with parameters $delay[f]$, $T2D$, and $\mathcal{D}$. Continue as before, sending *setup* messages along the shortest path from $z$ to $t_2$, until finally $t_2$ has been reached and has executed steps (*v*)-(*vii*). Then select $t_3$ as the node in $\mathcal{D} - \mathcal{T}$ which is closest to $\mathcal{T}$, and continue in this manner until all nodes in $\mathcal{D}$ have been added to $\mathcal{T}$.

We illustrate the method using Figure 4.7 and with $\Delta = 13$. Next to each arc $(i,j)$ is the arc cost $c_{ij}$. To simplify the exposition, assume that $d_{ij} = c_{ij}$ for all $(i,j) \in \mathcal{A}$; the algorithm of [54] allows $c_{ij}$ and $d_{ij}$ to be specified independently. There are three receiver nodes, $c$, $g$, and $p$. We initialize $\mathcal{T} = \{s\}$. $(i)$ At $s$ we initialize $T2D[p].cost = 5$, $T2D[p].treeNode = s$,

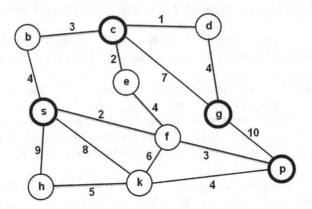

**Fig. 4.7** Delay constrained tree

$T2D[p].inTree = 0$, $T2D[c].cost = 7$, $T2D[c].treeNode = s$, $T2D[c].inTree = 0$, and $T2D[g].cost = 12$, $T2D[g].treeNode = s$, $T2D[g].inTree = 0$. Since the cost to $p$ is lowest, the shortest path from $s$ to $p$ is added to $\mathcal{T}$, and $s$ sends a *setup* message towards $p$. Since $f$ is the next node on the shortest path from $s$ to $p$, and at $s$ we have $RT_s[f].delay = 2$, we set $delay[f] = 2$ and include $delay[f]$, $T2D$, and $\mathcal{D}$ in the *setup* message to $f$.

Node $f$ determines that for $t = c$ and $t = g$ we have $RT_f[t].cost < T2D[t].cost$ and $delay[f] + RT_f[t].delay \leq \Delta$, and thus $f$ makes the updates $T2D[t].cost = RT_f[t].cost$ and $T2D[t].treenode = f$ for $t = c$ and $t = g$. Specifically, $f$ sets $T2D[c].cost = 6$ and $T2D[g].cost = 11$. The next node on the path from $f$ to $p$ is $p$ itself. The receiver nodes $c$ and $g$ are not yet in $\mathcal{T}$, and $p$ determines that $delay[p] + RT_p[c].delay = 5 + 9 > \Delta$, so no update to $T2D$ is made for $c$ by $p$. Similarly, $delay[p] + RT_p[g].delay = 5 + 10 > \Delta$, so no update to $T2D$ is made for $g$ by $p$.

Since $T2D[c].cost < T2D[g].cost$ (i.e., $6 < 11$), $c$ is selected as the next receiver node. For $c$ we have $T2D[c].treeNode = f$, so $f$ is a branching node. The shortest path from $c$ to $f$ is added to $\mathcal{T}$. Now a *fork* message is sent to $f$, which sends a *setup* message to $e$, the next node on the shortest path from $f$ to $c$. Node $e$ determines that for $g$ (the only receiver node not yet in $\mathcal{T}$), we have $RT_e[g].cost < T2D[g].cost$ (i.e., $7 < 11$) and $delay[e] + RT_e[g].delay < \Delta$ (i.e., $6 + 7 \leq \Delta$). So $e$ sets $T2D[g].cost = 7$ and $T2D[g].treenode = e$. The next node on the shortest path from $e$ to $c$ is $c$ itself, and $c$ determines that $RT_c[g].cost < T2D[g].cost$ (i.e., $5 < 7$) and $delay[c] + RT_c[g].delay < \Delta$ (i.e., $8 + 5 \leq \Delta$) so $c$ sets $T2D[g].cost = 5$ and $T2D[g].treenode = c$.

Now $g$ is selected at the final receiver node, and the shortest path from $c$ to $g$ is added to $\mathcal{T}$. Since all nodes in $\mathcal{D}$ have been added to the tree, we are done.

Huang and Lee [51] discovered that the network generated by Jia's method can contain cycles, and showed that with additional cycle processing and tree pruning steps, the modified method is (under Jia's two assumptions) guaranteed to produce a tree satisfying the delay constraint.

## 4.6 Ant-Based Method for Wireless MANETs

The 2008 method of Shen, Li, Jaikaeo, and Sridhara [99] utilizes *swarm intelligence* techniques in a distributed heuristic for dynamically building a delay constrained Steiner tree that minimizes the total transmission power in a wireless network. Let $(\mathcal{N}, \mathcal{A})$ be an undirected wireless network, where there is an arc $a \in \mathcal{A}$ between two nodes if they have sufficient power to communicate with each other. Since a wireless node broadcasts packets which can be received by any node within a radius determined by the transmission power (as opposed to wired networks which must duplicate packets on outgoing interfaces), the cost and delay functions in the problem formulation are defined over the set of nodes, not the set of arcs. The cost of each node depends on the transmission power assigned by the method to the node. If node $i$ forwards packets to $j$, then $i$ must have power at least $r_{ij}$ which depends on various parameters, including the distance between $i$ and $j$. If $i$ forwards packets to all nodes in some set $J$, then $i$ must have power at least $\max_{j \in J} r_{ij}$.

Let $c_p(n)$ be the cost of node $n$ when its power assignment is $p$. Let $d(n)$ be the transmission delay associated with using node $n$; the delay $d(n)$ is assumed to be independent of the power assigned to $n$ (this will be true, e.g., if delay measures hop count). Let $s$ be the source node, and let $\mathcal{D} \subseteq \mathcal{N} - \{s\}$ be the set of receiver nodes. A *forwarding set* $\mathcal{F} \subset \mathcal{N}$ is a set of nodes that provide a path from $s$ to all the receiver nodes; the forwarding set includes $s$ and may include receiver nodes which also act as forwarders. Since there are no physical arcs, each $t \in \mathcal{D}$ joins a unique forwarding node, which in turn joins another forwarding node, until the join reaches $s$. The problem is to determine a forwarding set $\mathcal{F}$, and a power assignment to each node in $\mathcal{F}$, that minimizes the objective function $\sum_{n \in \mathcal{F}} c_p(n)$, The tree must also satisfy the constraint that the delay from $s$ to each $t \in \mathcal{D}$ cannot exceed $\Delta$, for some specified bound $\Delta$.

The tree $\mathcal{T}$ is initialized when each $t \in \mathcal{D}$ computes a shortest path to $s$, where the cost of arc $(i, j) \in \mathcal{A}$ is proportional to the transmission power required for nodes $i$ and $j$ to communicate. Associated with each forwarding node $n$ in $\mathcal{T}$ is the number $height(n)$, which is defined as the highest node identifier of the branch of $\mathcal{T}$ rooted at $n$, and where $height(s) = \infty$ by definition. Once $\mathcal{T}$ is initialized, the method determines if these shortest paths can be merged by incorporating more forwarding nodes with lower power assignments. Employing the swarming idea, each receiver node periodically originates *forward ant* control messages. Two types of forward ant messages are used: ($i$) deterministic messages that are sent to the adjacent node $n$ for which the *pheromone intensity* (node desirability) is highest, and ($ii$) random ants, which explore other paths. Suppose the forward ant message from receiver node $t$ reaches a forwarding node $n$ on the tree for which $height(n) > height(t)$. If the total delay from $s$ to $t$ via $n$ does not exceed $\Delta$, then the forward ant becomes a *backward ant* and returns to $t$, while increasing the pheromone intensity of nodes along the way, in order to attract *join* requests to the new path. The requirement $height(n) > height(t)$ prevents a race condition in which receiver nodes attempt to merge onto each other's forwarding paths and end up disconnected from the source.

Receiver $t$ then recomputes its next hop by issuing a join to the neighboring node $x$ whose pheromone intensity is highest. The pheromone intensities decay over time, so that changing network information (e.g., a node failure) will be reflected in new intensities. In turn, node $x$ issues its own join, and this process continues until the join reaches the source. This is illustrated in Figure 4.8. The initial forwarding set is $\mathcal{F} = \{S, A, B, C, E\}$. When a for-

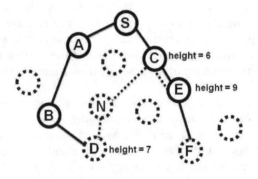

path of forward/backward ant •••••••••••

**Fig. 4.8** Ant based routing

ward ant from $D$ encounters $N$, it does not turn into a backward ant since $N \notin \mathcal{F}$. When it encounters $C \in \mathcal{F}$, it does not turn into a backward ant since $height(C) < height(D)$. When it encounters $E$, if the delay from $D$ to $S$ via $E$ does not exceed $\Delta$, it does turn into a backward ant, and increases the pheromone intensity of nodes $E$, $C$, and $N$.

# Chapter 5
# Inter-domain and Two-level Multicast

The *scope* (region of applicability) of an Interior Gateway Protocol (IGP), such as OSPF or ISIS, is a *domain*. The scope often corresponds to an AS, or it might be defined by the need to keep the number of nodes within the scale supported by the IGP protocol. An *Exterior Gateway Protocol*, typically BGP, interconnects two or more domains. Inter-domain multicast is required by an ISP, and is also required by a service provider whose customers span more than one IGP domain. In particular, when the set of source and receiver nodes for a multicast group spans more than one domain, inter-domain multicast routing is required.

Inter-domain multicast has two problems to solve. The first problem is how to reach a source that is in a different domain than a receiver, and the desired unicast and multicast topologies are not identical. This motivates the discussion of MP-BGP in Section 5.1 below.

The second problem is how to connect PIM-SM domains served by different RPs. For example, consider the case of two ISPs, each with its own AS. PIM-SM by itself does not allow more than one active RP for a given multicast group $g$. So if receivers in both ASs want to join $g$, the two ISPs must agree on whose RP will be used. In practice, such collaboration is unrealistic, since each ISP wants administrative control of its own RPs. MSDP solves this problem, by letting nodes in each domain know about active multicast sources in the other domain (MSDP was introduced in Section 3.8.4 on Anycast RP). This motivates the discussion of MSDP in Section 5.2 below.

Noting that inter-domain multicast is a kind of two-level routing (within a domain at the lower level, and between domains at the higher level), we also examine two other methods for two-level multicast routing: a method based on the Yao graph and DVMRP.

## 5.1 Multiprotocol BGP

BGP [102] is the de facto standard for inter-domain routing. It is used, e.g., by ISPs to peer (inter-operate) with other service providers, and also by service providers who administer more than one AS. *Multiprotocol BGP* (MP-BGP) is required when a multicast group spans multiple ASs, and when the desired multicast topology differs from the unicast topology. (MP-BGP can also be used inside a single AS when the desired unicast and multicast topologies differ.) MP-BGP was defined in 2000 by IETF RFC 2283, and updated in 2011 as an IETF proposed standard [11].

MP-BGP can distinguish between unicast routing and multicast routing by means of the *MP_REACH_NLRI* field in an MP-BGP message, where *NLRI* means *Network Layer Reachability Information*. The NLRI field for a reachable destination (e.g., a host or subnet) is the IP address prefix for the destination. For example, the NLRI for a subnet might be 10.2.8.0/22. With MP-BGP, BGP can provide both a unicast routing table and a multicast routing table. MP-BGP does not itself build multicast trees; rather, it is used to exchange routing information used by other protocols, such as PIM, to build trees and forward multicast traffic.

To distinguish between different address families, MP-BGP uses an *Address Family Identifier* (AFI). For example, AFI 1 is for IPv4 and AFI 2 is for IPv6. Within each AFI are sub-AFIs. For example, within AFI 1, information exchanged for multicast RPF calculations has sub-AFI 2.

Figure 5.1, adapted from [112], illustrates the use of MP-BGP. Router *B*,

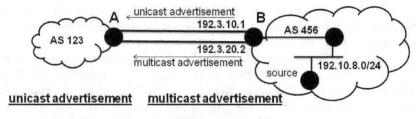

**Fig. 5.1** MP-BGP

in AS 456, has a single BGP session with router *A* in AS 123. Over this single BGP session, *B* advertises to *A* that the subnet 192.10.8/24 is reachable in AS 456, and to reach this address, *A* should send packets to the BGP next hop of 192.3.10.1. Over the same BGP session, *B* uses *MP_REACH_NLRI* with *AFI=1* and *sub-AFI=2* to advertise the multicast source 192.10.8/24, but the BGP next hop is now 192.3.20.2. This means that if *A* sends a PIM join to *B* (so that *A* can receive this multicast stream), then the PIM join from *A* must be sent to 192.3.20.2, and not to 192.3.10.1.

MP-BGP can be used in conjunction with SSM (Section 3.7.4) to build inter-domain multicast trees. Suppose a host subtending node *n* wants to receive the $(s, g)$ stream, and *g* is in the SSM address range. Then the host sends an IGMP Version 3 membership report to *n*, and then *n* sends a PIM join towards *s*. When a PIM join needs to cross from domain *X* to domain *Y*, the *MP_REACH_NLRI* information tells a border node of *X* the address in $Y$ to which the join should be sent. (A *border node* of a domain is a node in that domain that is directly connected to a node in a different domain.)

## 5.2 Multicast Source Distribution Protocol

*Multicast Source Distribution Protocol* (MSDP) was developed so that ISPs exchanging multicast could peer with each other. It is also used by service providers whose multicast networks span more than one domain. Consider a set of domains (ASs), each using PIM-SM as the multicast protocol. We allow the possibility that a domain might utilize multiple RPs, e.g., if anycast RP is used. To implement MSDP, we first require MSDP peering sessions to be established between pairs of RPs, between RPs and AS border routers, and possibly between redundant RPs serving the same multicast groups. There must be sufficiently many peering sessions to create a path of such sessions between any two RPs in different ASs. When a set of MSDP peers is fully meshed (i.e., each pair of RPs is linked by an MSDP session), the set is called a *mesh group*. For example, in Figure 5.2, the RPs do not form a mesh group, but there is a path of MSDP peers between any two RPs.

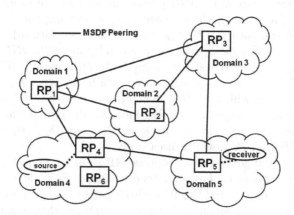

**Fig. 5.2** MSDP

Suppose an RP in a domain learns, through the standard PIM mechanisms, of a source host $s$ in that domain for group $g$. Then the RP encapsulates the first data packet from $s$ for $g$ in a *source active* (SA) message, which contains the identity of the source domain. What happens next depends if the RP is part of a mesh group. If it is, then the SA message is sent to all other members of the mesh group. These other members do not resend the message within this mesh group, but do send the message to any other MSDP peers which are not members of the mesh group. If the RP is not part of a mesh group, the SA message is sent to all the MSDP peers of the RP.

To prevent looping of MSDP messages, a modified RPF check is used by any node receiving an SA message. Consider a node $n$ which receives, from node $x$, an SA message that originated from a source host $s$ behind node $y$. A set of six rules [59] determine the *peer-RPF neighbor* of $n$ (i.e., the MSDP peer on the path from $n$ towards the source), and whether $n$ will accept or drop the message. These rules are applied in order, and the first rule that applies is used. One rule has already been noted: if $n$ and $x$ are in the same mesh group, then $n$ accepts the message from $x$. (Thus using a mesh group reduces SA flooding and simplifies the peer-RPF check.) Another rule is that if $x = y$, the message is accepted (i.e., accept the message if it is directly received from the source node). The last rule we mention applies to the case where $n$ and $y$ are in different ASs, where $n$ and $x$ and are in the same AS, and $x$ is an *autonomous system border router*, i.e., a node in one AS directly connected to a node in another AS. If $x$ is the BGP next hop from $n$ to $y$ (i.e., if $n$ learned about $y$ from $x$), then $x$ is the peer-RPF neighbor, and its SA messages are accepted by $n$.

We illustrate this using Figure 5.2. Assume that BGP routing has determined the best path from $D_1$ and $D_5$ to the source domain $D_4$ is a direct path (traversing no via domains), the best path from $D_2$ to $D_4$ is via $D_1$, and the best path from $D_3$ to $D_4$ is via $D_5$. Suppose $RP_4$ learns of the source in Domain 4 for group $g$. Then $RP_4$ sends an SA message to its peers $RP_1$, $RP_5$, and $RP_6$, who accept the message, since it was originated by the sending RP. $RP_6$ has no other MSDP peers, so it does not forward the message. $RP_1$ will send the SA message to its peers $RP_2$ and $RP_3$. $RP_2$ will accept the SA message, since the best path from $D_2$ to $D_4$ is via $D_1$. However, $RP_3$ will discard the SA message, since its best path to $D_4$ is via $D_5$, and none of the other RPF rules will cause $RP_3$ to accept the message. $RP_2$ forwards the message to $RP_3$, which discards it. $RP_5$ forwards the message to $RP_3$, which accepts it, since the best path from $D_3$ to $D_4$ is via $D_5$. $RP_3$ forwards the message to $RP_1$ and $RP_2$, which discard it.

Returning to the general case, suppose RP $n$, in domain $D_n$, receives an SA message about a source host $s$ subtending the source node $n(s)$ in another domain $D_s$ for group $g$, and the RPF check passes. If $n$ has no $(\star, g)$ state, then no receiver host in domain $D_n$ joined $g$, and hence there is no shared tree for $(\star, g)$ rooted at $n$. If $n$ has $(\star, g)$ state, then some receiver host interested in $g$ has joined to $n$, so $n$ creates $(s, g)$ state and sends an $(s, g)$ join towards $s$, in order to join the shortest path tree rooted at $n(s)$. Note that this tree will in general span multiple domains, and therefore both intra-domain and inter-domain protocols (e.g., OSPF and MP-BGP) may be required for this join. If $n$ has $(\star, g)$ state, then $n$ must also ensure that hosts in its domain $D_n$ know of $(s, g)$. So $n$ decapsulates the SA message and sends the packet down the $(\star, g)$ shared tree in $D_n$ rooted at $n$.

When a *last hop router* (the router on the tree that is closest to some receiver host) receives the decapsulated packet, it may itself join the $(s, g)$ shortest path tree rooted at $n(s)$. For example, referring to Figure 5.3, suppose host $r$ subtending the last hop router $z$ in $D_3$ has joined $RP_3$ to receive $g$. Then the $(s, g)$ stream, if its bandwidth does not exceed the SPT-threshold,

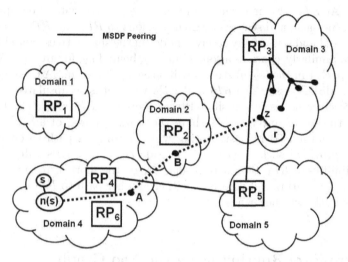

**Fig. 5.3** SPT join with MSDP

will take the path $s \rightarrow n(s) \rightarrow RP_4 \rightarrow RP_5 \rightarrow RP_3$, and then down the shared tree rooted at $RP_3$ to $r$, as illustrated by the solid lines. If the bandwidth of the stream does exceed the SPT-threshold, node $z$ will directly join the $(s, g)$ shortest path tree rooted at $n(s)$, so the $(s, g)$ stream takes the path $s \rightarrow n(s) \rightarrow A \rightarrow B \rightarrow z \rightarrow r$, as illustrated by the dotted lines.

MSDP allows load sharing by multiple RPs within a given PIM-SM multicast domain; this technique is called *anycast RP*, which was introduced in Section 3.8.4. Consider Figure 5.4. There are three RPs. With anycast RP,

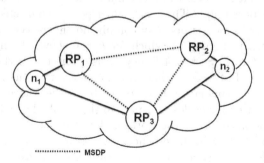

**Fig. 5.4** Anycast RP

we give the *same* unicast address to each of the RPs. Since a node uses unicast routing to send a PIM join or register message to the RP, a PIM join or register from $n_1$ will reach only $RP_1$, since it is closer to $n_1$ than is $RP_2$ or $RP_3$. Similarly, a join or PIM register from $n_2$ will reach only $RP_2$, since it is closer to $n_2$ than is $RP_1$ or $RP_3$.

Referring again to Figure 5.4, consider a source host $s_1$, behind $n_1$, for group $g$. At $RP_1$, the first packet in the $(s_1, g)$ stream will be encapsulated in an SA message, and sent over MSDP sessions to $RP_2$ and $RP_3$, where it is decapsulated and sent to any receiver nodes in the shared trees rooted at $RP_2$ and $RP_3$. Similarly, consider a source host $s_2$, behind $n_2$, for group $g$. At $RP_2$, the first packet in the $(s_2, g)$ stream will be encapsulated in SA messages, and sent over MSDP sessions to $RP_1$ and $RP_3$, where it is decapsulated and sent to any receiver nodes in the shared trees rooted at $RP_1$ and $RP_3$. Thus, in general, the set of anycast RPs share the work required of an RP. If $RP_1$ fails, then normal IP unicast routing will determine a path to either $RP_2$ or $RP_3$ (provided such a path exists) and these other RPs will share the responsibilities of $RP_1$. Similarly, if some arc on the shortest path from $n_1$ to $RP_1$ fails, IP routing will determine either another path to $RP_1$ or a path to a different RP, whichever has lower cost.

## 5.3 Two-level Routing using the Yao Graph

A two-level multicast routing method was proposed in 2008 by Lua, Zhou, Crowcroft, and Van Mieghem [67]. The multicast domain is subdivided into clusters. Within each cluster are the level-0 (lowest level) *peer nodes* and one or more level-1 *super-peer* nodes. Each peer is connected to its closest super-peer. The super-peer logical connectivity is established using the Yao graph [114] to partition two-dimensional Euclidean space $\mathcal{R}^2$. For a fixed positive integer $k$ ($k = 6$ is used in [67]) and for each super-peer $n$, the space around $n$ is partitioned into $k$ sectors, where each sector is an unbounded cone with angle $2\pi/k$ and with $n$ as a vertex. For $i = 1, 2, \cdots, k$, let $z_i(n)$ be that super-peer in sector $i$ that is closest to $n$. The undirected arc $(n, z_i(n))$ becomes an arc in the super-peer graph. Thus each super-peer is connected to at most $k$ other super-peers (some sectors may contain no super-peers other than $n$). This is illustrated in Figure 5.5 which for $k = 6$ shows the nodes to which node $n$ is connected. In Sectors 1-5, node $n$ is connected to that node in the sector closest to $n$. In Sector 6, there are no nodes other than $n$. This figure shows only the arcs emanating from $n$. To build the Yao graph, the same procedure is executed for *each* super-peer in the graph. The advantage of the Yao graph for mobile or ad hoc networks is that it uses only local information and is easily updated when there is network churn (e.g., nodes fail).

Routing at the level-1 super-peers layer uses *local compass routing*, in which a node $n$ selects the next hop based only on the destination $d$ and the nodes

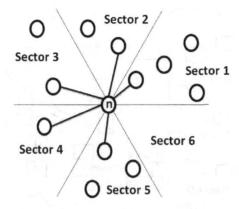

**Fig. 5.5** Yao graph

in the Yao graph adjacent to $n$. For each node $x$ adjacent to $n$ in the Yao graph, the angle $\angle xnd$ is computed, and the $x$ yielding the smallest angle is chosen as the next hop. A variant of this scheme is *random local compass routing*, in which $x_1$ is chosen to be node above the line through $n$ and $d$ which minimizes the angle $\angle x_1nd$, and $x_2$ is chosen to be node below the line through $n$ and $d$ which minimizes the angle $\angle x_2nd$; then $n$ randomly chooses $x_1$ or $x_2$ as the next hop.

Inter-cluster routing (at the super-peers level) from source node $s$ is accomplished by building source trees. When a source peer node $s$ wants to send a packet to group $g$, it sends the packet (using unicast routing) to the closest super-peer $n(s)$; the super-peer $n(s)$ must also join $g$ if it has not already done so. A super-peer $n$ receiving a packet for group $g$ does an RPF check to see if the packet arrived on the arc that $n$ would use to route to $s$; if the check passes, the packet is forwarded on all links from $n$ to other super-peers. Pruning can be used to remove branches of the tree with no downstream group members.

## 5.4 Two-level Routing using DVMRP

Thyagarajan and Deering [106] proposed in 1995 a method that partitions a multicast domain into regions. A unique region identifier (ID) is assigned to each region. Each region contains level-1 (L1) nodes responsible for forwarding multicast traffic within a region; each L1 node belongs to a single region. Level-2 (L2) boundary nodes are used to forward traffic between regions; an L2 node connecting regions $A$ and $B$ also provides L1 functionality to regions $A$ and $B$. This is illustrated in Figure 5.6, where there are two L1 nodes in Region $A$, two in Region $B$, four in $C$, and one in $D$. One L2 node

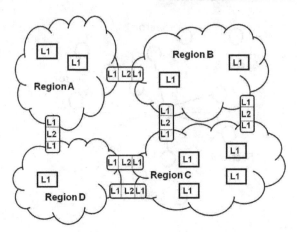

**Fig. 5.6** Level 1 and level 2 routers

interconnects Regions $A$ and $B$, one interconnects $A$ and $D$, two interconnect $B$ and $C$, and two interconnect $C$ and $D$, where each L2 node also provides L1 functionality in the two regions it interconnects.

The L1 nodes in a region are assumed to run the same standard multicast protocol (which, at the time of this proposal, included DVMRP and MOSPF). Different regions may run different intra-region protocols. Consider a packet for group $g$ originated in a given region $R$. The intra-region protocol is used to forward the packet to all receiver nodes for $g$ in $R$, and, by default, to *all* the L2 nodes attached to $R$.

The L2 nodes perform L2 inter-region routing using DVMRP, which creates a tree for $g$, spanning all the regions, and rooted at the source node in the source region (the source node, for this computation, is treated as an L2 node). For example, suppose in Figure 5.7 that there are receiver hosts for $g$ in Region $C$ subtending the L1 nodes $L1c$, $L1d$, $L1e$, and $L1f$. Suppose the DVMRP tree is given by the heavy solid lines (which represent a path in a region, and not necessarily a single link). There are three L2 nodes attached to Region $C$ used in the L2 tree: $L2a$, $L2b$, and $L2d$. Each of the four nodes L1 nodes in Region $C$ must determine from which of those three L2 nodes they will receive packets for $g$. To accomplish this, each L2 node floods within its two attached domains a default cost. Using these default costs, each L1 node determines the closest attached L2 node. This yields, in each region, a partition of the L1 nodes, with the L1 nodes in each partition forming a subtree rooted at an L2 node. For example, in Figure 5.7, in Region $C$ the nodes $L1c$, $L1d$, and $L1e$ are in the subtree rooted at $L2a$; the nodes $L1f$, $L1g$, and $L1h$ are in the subtree rooted at $L2b$; and the nodes $L1w$ and $L1x$ are in the subtree rooted at $L2d$ (no L1 nodes in region $C$ connect to $L2c$ for this $g$).

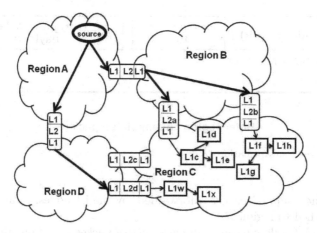

**Fig. 5.7** Level 2 tree created by DVMRP

For each region $R$, all the L2 nodes attached (i.e., directly connected) to region $R$ form an L2 multicast group $g(R)$; the number of L2 multicast groups is thus equal to the number of regions. For an L2 multicast group $g(R)$, a source is an L2 node attached to $R$, and the receivers are all the other L2 nodes attached to $R$.

A Region Membership Report (RMR) is used to inform all L2 nodes attached to $R$ about hosts in $R$ interested in receiving traffic for $g$. An L1 node $x$ in $R$ sends an RMR periodically, and upon triggered events, e.g., when the first host subtending $x$ joins $g$, or when the last host subtending $x$ leaves $g$. This is analogous to the use of IGMP by hosts to signal group membership. The L2 multicast group $g(R)$ is used to distribute the RMR to all the L2 nodes attached to $R$.

If an L2 node $n$ attached to $R$ learns through RMR reports that there are no receivers for $g$ in $R$, then $n$ issues a DVMRP prune to its upstream L2 node. If sometime later a receiver for $g$ does appear in $R$, then $n$ can issue a DVMRP graft message for $g$ to its upstream L2 node to get reconnected to the tree for $g$.

Let $R_1(n)$ and $R_2(n)$ be the two regions connected by the L2 node $n$. Node $n$ knows the ID of all the regions, and, by RMR reports, $n$ also knows the addresses of all the receiver hosts in the regions $R_1(n)$ and $R_2(n)$. When an L2 node $n$ receives a *non-encapsulated* packet for $g$, the following happens.

($i$) Node $n$ checks that the source of the packet is in one of its attached regions $R_1(n)$ or $R_2(n)$; if not, the packet is discarded. Suppose, without loss of generality, that the source is in region $\bar{R} = R_1(n)$.

($ii$) Node $n$ encapsulates the packet with a tag specifying the source region, and with an outer label specifying the L2 node $n$, and the group $g(\bar{R})$. The pair $(n, g(\bar{R}))$ is the (source, group) for the L2 multicast group. This is illustrated in Figure 5.8.

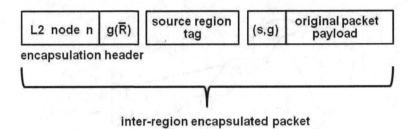

**inter-region encapsulated packet**

**Fig. 5.8** Level 2 encapsulation

($iii$) The encapsulated packet is forwarded, by the L1 nodes, to all other L2 nodes attached to region $\bar{R}$.

When an L2 node $n$ receives an *encapsulated* packet for $g$ from an attached region, say $R_1(n)$, the following happens.

($i$) Node $n$ decapsulates the packet and an RPF check is performed, using the originating region tag, to see if the packet arrived at $n$ on the shortest path to the source region; if not, the packet is discarded.

($ii$) If there are any receivers for $g$ in the regions $R_1(n)$ or $R_2(n)$ attached to $n$, then a copy of the original packet is injected into the region, where the L1 nodes in that region forward the packet to the receivers.

($iii$) Suppose there are regions downstream of $n$ (i.e., downstream of $n$ with respect to the L2 inter-region tree created by DVMRP for this $g$) with receivers. Such downstream regions can be identified using the source region tag and the group address of the original packet. For each downstream region with receivers interested in $g$, the packet is forwarded across $R_2(n)$ to the L2 nodes attached to $R_2(n)$ that are on the path to the downstream regions; at these L2 nodes the above encapsulation process is applied.

This method can be extended to hierarchical multicast with $h$ levels of hierarchy (the above discussion assumed $h = 2$). For example, with $h = 3$, we can cluster regions into super-regions interconnected by level-3 routers, and apply the encapsulating scheme recursively. Although this method was an early advance in inter-domain multicast, it never enjoyed wide deployment, since MSDP became the method of choice to link multicast domains.

# Chapter 6
# Aggregate Multicast Trees

We have seen how a shared tree, e.g., as created by BiDir, reduces multicast state by requiring only $(\star, g)$ state per node, rather than $(s, g)$ state per node for a source tree. The subject of state reduction has enjoyed considerable attention, and various techniques, such as tunnelling, have been proposed to reduce state ([40], [74]). Even greater state reduction can be achieved by forcing multiple groups to use the same tree, in which case the tree is called an *aggregate tree*. With aggregate trees, *backbone* nodes (with no subtending source or receiver hosts) need only keep state per aggregate tree, and not per group. *Edge* nodes (with subtending source or receiver hosts) still need to keep state on a per group basis. Various approaches to aggregated multicast are described in [13] and [107].

Fei, Cui, Gerla, and Faloutsos [40] describe an aggregate tree scheme for inter-domain multicast. Referring to Figure 6.1, suppose Domain 1 is the *backbone network* (interconnecting the backbone nodes) of an ISP, and the other domains attach to Domain 1. Consider first a group $g_1$ with a source

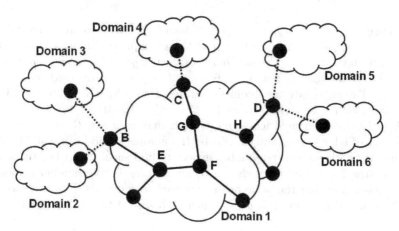

**Fig. 6.1** Tree aggregation

host in Domain 6, and receiver hosts in Domains 3, 4, and 5. The shared tree for this group will impose $(\star, g_1)$ state on the seven nodes $B$, $C$, $D$, $E$, $F$, $G$, $H$. Now consider a second group $g_2$ that also has a source host in Domain 6, and receiver hosts in Domains 2, 4, and 5. The shared tree for this group will impose $(\star, g_2)$ state on the same set of seven nodes. Since within the backbone (Domain 1) the same multicast tree is used, we can use a single aggregate tree that spans these seven nodes. Thus, in Figure 6.1, for the aggregate tree we must maintain one aggregate state on $B$, $C$, $D$, $E$, $F$, $G$, $H$. However, we must still maintain per group state ($g_1$ and $g_2$) on the edge nodes $B$, $C$, and $D$; this is required in order for these nodes to correctly route packets to the receiver hosts. For example, $B$ requires both $g_1$ and $g_2$ state in order to decide whether to send packets to Domain 2 or Domain 3. Thus we must carry per group $(\star, g)$ state on each packet as it traverses the backbone, so that this information is available to the edge nodes. This state is called *irreducible* since it must be maintained by edge nodes. One way to implement such a scheme is to use MPLS, where an inner label carries $(\star, g)$ and the outer label identifies the aggregate tree.

Define the *total state* in the network $(\mathcal{N}, \mathcal{A})$ as $\sum_{n \in \mathcal{N}} state(n)$, where $state(n)$ is the total number of aggregate and $(\star, g)$ states on node $n$. Using the method of [40], we can estimate the reduction in total state obtained by using aggregate trees versus not using aggregation (i.e., using a distinct shared tree for each group). Let $S_a$ be the total state using aggregate trees, and let $S_0$ be the total state without aggregation. A simple measure of the state reduction with aggregate trees is $R = (S_0 - S_a)/S_0 = 1 - S_a/S_0$. Since some of the state is irreducible, a better measure of the state reduction is obtained by letting $S_{ir}$ be the irreducible state, and defining

$$R_{ir} = \frac{(S_0 - S_{ir}) - (S_a - S_{ir})}{S_0 - S_{ir}} = 1 - \frac{S_a - S_{ir}}{S_0 - S_{ir}}.$$

We have $0 \leq R_{ir} \leq 1$, and larger values of $R_{ir}$ mean more reduction in state by using an aggregate tree. For the example of Figure 6.1, without aggregation the trees in the backbone for $g_1$ and $g_2$ each require 7 states, so $S_0 = 14$. With aggregation, $S_{ir} = 6$, since 3 edge nodes are used by $g_1$, and 3 by $g_2$. The aggregate tree requires 7 states. Hence $S_a = 7 + (2)(3) = 13$, so $R = 1 - 13/14 = 1/14$, and $R_{ir} = 1 - (13 - 6)/(14 - 6) = 1/8$.

Let $\mathcal{G}$ be a set of multicast groups such that each $g \in \mathcal{G}$ uses the exact same set of edge nodes, and let $N_E$ be the number of nodes in this set. (In a large network, we can reasonably expect to find such a set.) Let the same shared tree $\mathcal{T}$ be used for each $g \in \mathcal{G}$, and let $N_T$ be the number of nodes in $\mathcal{T}$. Assuming that the same procedure used to construct $\mathcal{T}$ is also used to build a shared tree for each $g \in \mathcal{G}$, then each shared tree will also have $N_T$ nodes.

Using a distinct shared tree for each group, the total state $S_0$ is given by $S_0 = G N_T$, where $G = |\mathcal{G}|$. Using one aggregate tree, the total state is $S_a = N_T + G N_E$, since each of the $N_T$ nodes in the aggregate tree gets aggregate state, and each of the $N_E$ edge nodes gets $(\star, g)$ state for $g \in \mathcal{G}$. Hence $R = 1 - (N_T + G N_E)/G N_T$. Since the state for the $G$ groups on the $N_E$ edge nodes is irreducible, we have

$$R_{ir} = 1 - \frac{(N_T + GN_E) - GN_E}{GN_T - GN_E} = 1 - \frac{N_T}{G(N_T - N_E)}.$$

This expression quantifies the intuitive result that an aggregate tree is more effective when the number $G$ of groups being aggregated is large, and the number $N_E$ of edge nodes used by each group is small compared to the number $N_T$ of nodes in the shared tree. For example, referring again to the above example of Figure 6.1, suppose that, instead of just the two groups $g_1$ and $g_2$, we have $G$ groups, where each of the $G$ groups spans the same set of domains as $g_1$. Then $N_E = 3$ (three edge nodes in each group), $N_T = 7$ (seven nodes in the aggregate tree), $S_0 = 7G$, and $S_{ir} = 3G$. Thus $R = 1 - (7 + 3G)/7G \approx 4/7$ for large $G$, and $R_{ir} = 1 - 7/(7G - 3G) \approx 1$ for large $G$, which indicates the efficiency possible with aggregation.

## 6.1 Leaky Aggregate Trees

Consider an aggregate tree $T$ with node set $\mathcal{N}_T$, and consider a set $\mathcal{G}$ of multicast groups, where $\mathcal{N}_g$ is the set of receiver nodes used by $g \in \mathcal{G}$. We say that $T$ covers $g$ if $\mathcal{N}_g \subseteq \mathcal{N}_T$. If $T$ covers $g$, then $T$ can be used as a shared tree for group $g$. We do not allow $T$ to be used for $g$ if $\mathcal{N}_g \not\subseteq \mathcal{N}_T$. If $\mathcal{N}_g$ is a proper subset of $\mathcal{N}_T$ for some $g$, then *leakage* occurs, meaning that bandwidth is wasted delivering traffic for $g$ to a node behind which are no interested hosts. For example, in Figure 6.1, if $\mathcal{N}_T$ spans the edge nodes $B$, $C$ and $D$ but $\mathcal{N}_g$ spans only $B$ and $C$, then leakage occurs since traffic is unnecessarily delivered to $D$. While leakage is undesirable, some leakage may be acceptable in order to reduce multicast state.

Assume $T$ covers $g$ for $g \in \mathcal{G}$. Let $c(T)$ be the cost of $T$, where, as usual, the cost of a tree is the sum of the arc costs over all arcs in the tree. Let $T_g$ be a tree spanning the receiver nodes of $g$. Defining $G = |\mathcal{G}|$, the *average leakage* $L(T, \mathcal{G})$ incurred by using the aggregate tree $T$ rather than the set of individual trees $T_g$, $g \in \mathcal{G}$ is defined in [40] as

$$L(T, \mathcal{G}) = \frac{c(T) - (1/G) \sum_{g \in \mathcal{G}} c(T_g)}{(1/G) \sum_{g \in \mathcal{G}} c(T_g)}.$$

The larger $L(T, \mathcal{G})$ is, the more bandwidth is wasted using $T$ rather than using the group specific trees $T_g$ for $g \in \mathcal{G}$. This measure suggests a strategy to determine which groups to aggregate [40]. Let $\mathcal{G}$ be a set of multicast groups for which we are considering using one or more aggregate trees. Consider a set $\mathcal{S}$ which contains a set of aggregate trees that will be used to cover the groups in $\mathcal{G}$. For $T \in \mathcal{S}$, let $\mathcal{G}(T)$ be the set of groups in $\mathcal{G}$ which are covered by the tree $T$. Initially, $\mathcal{S} = \emptyset$. By *extending* a tree $T$ to yield a new tree $T'$

we mean adding branches to $\mathcal{T}$. Let $\delta$ be a given threshold parameter. The strategy is implemented in procedure *Aggregate_Trees*() below.

```
procedure Aggregate_Trees(𝒢)
1       initialize: 𝒮 = ∅;
2       for (g ∈ 𝒢) {
3           for (𝒯 ∈ 𝒮) {
4               if (𝒯 covers g and L(𝒯, 𝒢(𝒯) ∪ {g}) ≤ δ) set x(𝒯, g) = 1;
5               else set x(𝒯, g) = 0;
6           }
7           if (x(𝒯, g) = 0 for all 𝒯 ∈ 𝒮), do either (i) or (ii) {
8               (i) extend some 𝒯 ∈ 𝒮 to create a new tree 𝒯' covering g
                    such that L(𝒯', 𝒢(𝒯) ∪ {g}) ≤ δ;
9                   set 𝒢(𝒯') = 𝒢(𝒯) ∪ {g};
10              (ii) set 𝒢(𝒯_g) = {g} and add 𝒯_g to 𝒮;
11          }
12          else {
13              pick 𝒯* such that x(𝒯*, g) = 1 and
                    c(𝒯*) = min{c(𝒯) | x(𝒯, g) = 1};
14              set 𝒢(𝒯*) = 𝒢(𝒯*) ∪ {g};
15          }
16      }
```

**Fig. 6.2** *Aggregate_Trees*()

A significant practical issue that must be addressed with an approach such as procedure *Aggregate_Trees*() is that group membership is dynamic, so an aggregate tree covering a group $g$ may cease to cover $g$ when new receivers join the group.

## 6.2 Hierarchical Aggregation

Let $\mathcal{D}$ be a multicast domain with node set $\mathcal{N}$, and let $\mathcal{N}_g \subseteq \mathcal{N}$ be the set of source or receiver nodes used by group $g$. If for some $g_1$ and $g_2$ we have $\mathcal{N}_{g_1} = \mathcal{N}_{g_2}$ then the same shared tree can be used for $g_1$ and $g_2$. If the members of $g_1$ and $g_2$ are chosen randomly, the probability that $\mathcal{N}_{g_1} = \mathcal{N}_{g_2}$ is small, and this probability decreases with the size of $\mathcal{N}$. Therefore, if we divide $\mathcal{D}$ into two subdomains, the likelihood of two groups using the same set of nodes increases. With this motivation, Moulierac, Guitton, and Molnár [74] divide a domain into two or more subdomains, aggregate within each subdomain, and stitch the subdomains using tunnels.

A simple approach to dividing a domain is to first find the nodes $n_1$ and $n_2$ furthest apart; we initialize $\mathcal{D}_1 = \{n_1\}$ and $\mathcal{D}_2 = \{n_2\}$. At each subsequent iteration, the unassigned node nearest to $\mathcal{D}_1$ is assigned to $\mathcal{D}_1$ and the unassigned node nearest to $\mathcal{D}_2$ is assigned to $\mathcal{D}_2$. Subdomains can be further divided, if desired.

Subdomain $\mathcal{D}_i$ is controlled by a processor $p_i$, which knows the topology of $\mathcal{D}_i$ and the groups with members in $\mathcal{D}_i$. Each subdomain contains one or more *border* nodes, which are nodes that are directly connected to a node in another subdomain. Suppose border node $n \in \mathcal{D}_i$ receives, from some other subdomain, a join request for $g$. Then $p_i$ adds $n$ to the set of nodes used by $g$ within $\mathcal{D}_i$, and decides whether to use an existing aggregate tree for $g$ in $\mathcal{D}_i$ (e.g., if the bandwidth leakage does not exceed a given tolerance) or to build a new tree for $g$ in $\mathcal{D}_i$.

A central processor $p$ stitches together the subdomains for each group $g$. Each $p_i$ selects, for $g$, a border node $b_i(g)$ in its subdomain. Processor $p_i$ informs processor $p$ of the chosen $b_i(g)$, and $p$ creates a set of unicast tunnels to join these $b_i(g)$ nodes in a global tree for $g$. For example, suppose there are 8 subdomains, but that $g$ spans only $\mathcal{D}_i$, for $i = 1, 3, 4, 6$. Then $p$ might create the 3 bidirectional tunnels $(b_1(g), b_3(g))$, $(b_1(g), b_4(g))$, and $(b_1(g), b_6(g))$ to form a tree linking these 4 border nodes.

The method was simulated on a network with 201 routers and 10,000 multicast groups. For simplicity, each $p_i(g)$ was chosen randomly from among the border nodes in subdomain $i$ for $g$; more sophisticated schemes could be devised. Define *total state* to be the sum of ($i$) the total router state in all the routers and ($ii$) the state required for the inter-subdomain tunnels. Using 4 subdomains, a 16% reduction in total state was achieved relative to 1 undivided domain when no bandwidth leakage is allowed; with 20% bandwidth leakage allowed, the total state is reduced by 25%. Having 4 subdomains is almost always better than having only 1 or 2.

## 6.3 Aggregation of Forwarding State

A standard technique in unicast routing is the use of a summary address to represent a collection of more specific addresses. Since unicast addresses are often based upon geographical location, summary addresses can be created based on a geopolitical hierarchy [8], or a structured numbering plan [95]. For IP routing, *variable length subnet masking* (VLSM) [77] is used to represent many subsets by a single address prefix. Because multicast group addresses are not tied to a geopolitical hierarchy, it is commonly believed that multicast state on a router cannot be aggregated. Thaler and Handley [104] demonstrate that multicast state in the forwarding table can be aggregated (so that less memory is consumed by the forwarding table), even when multicast addresses and group members are randomly distributed. They consider only perfect aggregation (as opposed to leaky aggregation), so that no traffic is needlessly sent to a node.

The basis of their aggregation scheme is a paradigm shift from the traditional packet forwarding model (Section 1.5.1) that specifies, for each $(\star, g)$ or $(s, g)$ entry, an RPF interface and an outgoing interface list (OIL). In the traditional model, an incoming packet for $(\star, g)$ or $(s, g)$ will be accepted only if it arrives on the specified RPF interface; if so, the packet is replicated and sent out on each interface in the OIL. The paradigm shift is to instead associate with each interface an input filter and an output filter. For interface $i$, Let $F_i^I$ be the input filter and $F_i^O$ be the output filter. A filter has the output 1 (accept), 0 (reject), or "−" (don't care). When a packet arrives on interface $i$, it is dropped if $F_i^I = 0$; otherwise, the output filter of each other interface on the node is checked to see if the packet should be replicated and sent out that interface.

While their model applies to source trees, which require $(s, g)$ state, consider for simplicity the case of shared trees. We assume that a sparse mode protocol like PIM-SM is run, so that a node $n$ maintains $(\star, g)$ state only if there are receiver hosts for $g$ downstream of $n$. Consider a particular interface $i$, and consider the input filter $F_i^I$ (this choice was arbitrary, and we could have picked $F_i^O$). Let $\mathcal{G}$ be the address space, e.g., $\mathcal{G}$ is the set of multicast addresses from 224.0.0.0 to 239.255.255.255, and let $G = |\mathcal{G}|$ be the size of the space. The crux of this method is to partition $\mathcal{G}$ into a set of regions. Referring to Figure 6.3, in region $k$, the addresses span the range $a_k$ to $a_{k+1} - 1$, where $F_i^I(a_k) = 1$, and $F_i^I(a_{k+1} - 1) = 0$. Each region is

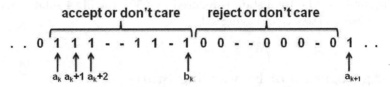

**Fig. 6.3** State aggregation

further divided into two sub-regions: for some $b_k$ such that $F_i^I(b_k) = 1$ and $F_i^I(b_k + 1) = 0$ we have $F_i^I(g) = 1$ or "−" for $g \in [a_k, b_k]$, and $F_i^I(g) = 0$ or "−" for $g \in [b_k + 1, a_{k+1} - 1]$. The demarcators $a_k$ and $b_k$ for subregion $k$ depend in general on the interface $i$ and whether we are considering the input or output filter for $i$. The significance of these two sub-regions of each region is that the behavior of $F_i^I$ in the range $[a_k, a_{k+1} - 1]$ can be summarized by the 3 demarcators $a_k$, $b_k$, and $a_{k+1}$. It is in this sense that we have aggregated the state. If $\mathcal{G}$ partitioned into $K$ regions, then the behavior of $F_i^I(g)$ for $g \in \mathcal{G}$ is summarized by at most $3K$ demarcators. Therefore the objective is to have $K$ as small as possible.

For simplicity of notation, we write $F$ rather than $F_i^I$. Assume that $F$ is a random variable defined on $\mathcal{G}$, and the values $\{F(g) \,|\, g \in \mathcal{G}\}$ are independent.

For $g \in \mathcal{G}$, let $p$ be the probability that $F(g) = 1$, let $q$ be the probability that $F(g) = 0$, and let $r = 1 - p - q$ (thus $r$ is the "don't care" probability). Then the expected number $K$ of regions is approximately $Gpq/(1 - r)$. The proof is as follows [104]. Define a success as the event that $F(g)$ is either 1 or "$-$"; the probability of a success is $1 - q$. The number of successes until a failure follows a geometric distribution with probability $q$. The expected number of successes until a failure, which is the expected length of the "accept or don't care" sub-region in Figure 6.3, is $\sum_{j=1}^{G} j(1 - q)^j q$. For large $G$, this sum is approximated by $\sum_{j=1}^{\infty} j(1 - q)^j q = (1 - q)/q$. Now re-define a success to be the event that $F(g)$ is either 0 or "$-$"; the probability of a success is now $1 - p$. The number of successes until a failure follows a geometric distribution with probability $p$, and the expected number of successes until a failure, which is the expected length of the "reject or don't care" sub-region in Figure 6.3, is approximately $\sum_{j=1}^{\infty} j(1 - p)^j p = (1 - p)/p$ for large $G$. The expected number of addresses that can be aggregated into a single region is therefore $[(1 - q)/q] + [(1 - p)/p] = (p + q - 2pq)/pq \approx (p + q)/pq = (1 - r)/pq$. Finally, since $G = |\mathcal{G}|$, the expected number $K$ of regions is approximately

$$K = G/[(1 - r)/pq] = Gpq/(1 - r). \tag{6.1}$$

Considering a given node $n$, define $\mathcal{G}_n$ to be the subset of $\mathcal{G}$ such that node $n$ has $(\star, g)$ state for $g \in \mathcal{G}_n$. Let $G_n = |\mathcal{G}_n|$. We consider one application of the formula $K = Gpq/(1 - r)$ for input filters, and one application for output filters; more applications are discussed in [104]. A *customer facing* interface is an interface connected to a host, as opposed to an interface connected to another node.

($i$) Consider the input filter associated with some customer facing interface on node $n$. Suppose the tree for each $g$ is a bidirectional shared tree, and suppose the addresses $g$ are uniformly distributed. Since no incoming interface check (i.e., no RPF check) is done, then $p = G_n/G$. Since also $r = 0$, then $q = 1 - p$ and (6.1) yields $K = G_n[1 - (G_n/G)]$ for this input filter. This expression is maximized when $G_n = G/2$. If $G_n << G$, then $K \approx G_n$, so no savings due to aggregation has been achieved. However, $G_n[1 - (G_n/G)]$ is a decreasing function of $G_n$ for $G_n > G/2$, and $K \to 0$ as $G_n \to G$.

($ii$) Consider the output filter associated with some interface $i$ on node $n$. Suppose the addresses $g$ are uniformly distributed, and suppose there are $J$ downstream interfaces at $n$ for each $g$. Suppose also that for each $g$ there are $M$ nodes downstream from $n$, and each of these $M$ nodes is equally likely to join on one of the $J$ downstream interfaces. Since all unused addresses are "don't care," then $r = 1 - (G_n/G)$. The probability $q$ that the output filter value for interface $i$ is 0 for some $g$ is the probability that $g \in \mathcal{G}_n$ and all $M$ downstream nodes join on some interface other than $i$; thus $q = (G_n/G)Q$, where $Q = [(J - 1)/J]^M$. Then $p = 1 - q - r = (G_n/G)(1 - Q)$ and (6.1) yields $K = G_n Q(1 - Q)$. Since $G_n Q(1 - Q) < G_n Q$ and $Q$ is a decreasing function of $M$, then $K$ decreases as $M$ increases. Hence aggregatability increases as the average number of downstream members of a group increases.

## 6.4 Subtree Aggregation

We have shown how aggregate trees can be used, at the cost of some leakage if the groups being aggregated do not span the same set of receiver nodes. Another approach is to search for a subtree that is common across the trees used by a set of multicast groups, and to replace the common subtrees by a single aggregate subtree. Since only exactly the same subtrees are combined, there is no leakage with this method. As shown by Ciurea [26], this scheme is easily implemented with MPLS P2MP trees (Section 3.10). In order to implement subtree aggregation, we use downstream, on demand, ordered label assignment [29]. *Downstream* assignment means that a downstream node $n$ generates a label to be used by its parent node for any packet destined for $n$. *On demand* means that $n$ generates a label only if $n$ has received a request to generate a label. *Ordered* means that label assignment proceeds in an orderly fashion from one end of the LSP to the other.

Ciurea's scheme is illustrated in Figure 6.4, where the receiver nodes for group $x$ subtend nodes $B$, $E$, $J$, and $L$, and the receiver nodes for group $y$ subtend, at the outset, nodes $C$ and $E$. Assume $A$, $D$, and $H$ are backbone nodes, and all other nodes are edge nodes, at which all labels are imposed or removed. Suppose we start out with separate MPLS P2MP trees for $x$

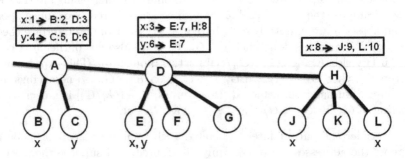

**Fig. 6.4** Initial configuration

and $y$. Assume we have the following rule: if the OILs at a node $n$ for two groups are identical, and if the upstream neighbor $p$ of $n$ is the same for both groups, then $p$ can forward packets to $n$ using the same label for both groups. Above each node in Figure 6.4 are the entries in the *Multicast Forwarding Information Base* (MFIB) for groups $x$ and $y$, e.g, if a group $x$ packet is received by $A$ with label 1, one copy is sent to $B$ with label 2, and one copy is sent to $D$ with label 3. (With MPLS, labels are locally assigned, but for clarity the figure uses globally unique labels. Also, in practice the MFIB would specify the interface to an adjacent node, rather than just the node identifier.) At $H$, if a group $x$ packet is received with label 8, it is sent to $J$ with label 9 and to $L$ with label 10.

Now suppose a host behind $J$ wants to join group $y$. Then $J$ must tell its parent $H$ to send down packets for $y$. So $J$ reuses the label 9 and tells $H$ to send group $y$ packets to $J$ with label 9. Node $H$ must also tell its parent $D$ to send down packets for $y$. However, since at $H$ the OILs for $x$ and $y$ are not identical, $H$ generates the new label 11 and tells $D$ to send group $y$ packets to $H$ with label 11. The updated MFIB entries at $H$ and $D$ for $y$ are shown in Figure 6.5. Now suppose a host behind $L$ wants to join group $y$. At this

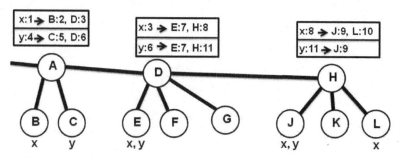

**Fig. 6.5** Host behind J joins group y

point, $H$ recognizes that its OILs for $x$ and $y$ are identical, combines them into a single OIL, and advertises the single label 8 to its upstream neighbor $D$, as as shown above $H$ in Figure 6.6. When $D$ receives the message to send

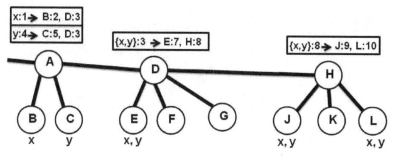

**Fig. 6.6** Host behind L joins group y

group $y$ packets to $H$ with label 8, $D$ recognizes that its OILs for groups $x$ and $y$ are now identical, and combines them into a single OIL, as shown above $D$ in Figure 6.6. At $A$ the two OILs are not identical, so they cannot be combined.

Finally, suppose the host behind $L$ leaves group $y$. Then $H$ must disaggregate, and create separate MFIB entries for $x$ and $y$, so the MFIB at $H$ reverts to the one shown in Figure 6.5. Similarly, $D$ must now disaggregate, and its MFIB reverts to the one shown in Figure 6.5. For this reason, [26] suggests that each node store previously generated upstream label assignments.

In this example, when the host behind $L$ leaves $y$, the label updates propagate only up to $D$. In general, with subnet aggregation, label updates for group $g$ will propagate all the way to the root of the P2MP tree for $g$.

# Chapter 7
# Multicast Virtual Private Networks

When a customer (e.g, a business, university, or branch of government) needs secure and reliable telecommunications, it can build its own network, or it can go to a *Service Provider* (SP), such as AT&T, who offers *Virtual Private Network* (VPN) services over its backbone network. The adjective *virtual* reflects the fact that the customer does not have to manage a backbone network, that the SP backbone routing procedures are transparent to the customer, and that multiple customers can use the SP backbone without impacting each other (if the SP backbone is properly engineered). A customer's VPN may handle only unicast traffic, or may handle both unicast and multicast traffic; in the latter case, the VPN is called a *multicast VPN* (mVPN). Since multicast VPNs are an extension of unicast VPNs, we first briefly describe how unicast VPNs work. For brevity, we consider only Layer 3 VPNs; see [49] for Layer 2 VPN methods.

## 7.1 Unicast VPNs

A VPN customer $C$ pays for two or more interfaces in the SP backbone network, and the VPN for $C$ provides the ability to route IP packets between the interfaces for $C$, just as if $C$ had built its own dedicated network. Since typically each customer contracts with the SP for a single VPN, we will for simplicity use *customer* and *VPN* synonymously, recognizing that some customers do order multiple VPNs from a given SP. Figure 7.1 illustrates a service provider VPN architecture. Each host for $C$ connects (not necessarily directly) to a *customer edge* (CE) router, typically owned and managed by $C$; for simplicity, in this figure only two CEs are shown. Each CE connects to one or more *provider edge* (PE) routers, owned and managed by the SP. Many different CE routers, belonging to different customers, can connect to a given PE. The PE routers have the responsibility for providing VPN features, such as the ability to provide different classes of service for different types of customer traffic, and the ability to terminate different types of customer interfaces (e.g., Ethernet or Packet over SONET). For simplicity, we assume each CE connects to a single PE. Each PE connects to one or more *backbone* routers in the SP network. Service provider backbone routers are called $P$ routers. A PE router might connect only to the nearest P router, or it might connect to two P routers, e.g., to provide an alternate route if one of the PE-

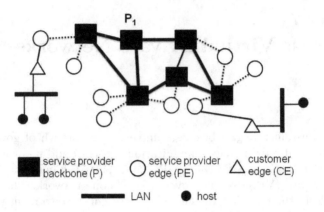

**Fig. 7.1** VPN architecture

P links fails. The P routers, typically running the IP and MPLS protocols, are workhorses designed to accept high speed interfaces and forward packets to other P routers and to subtending PE routers. (In Figure 7.1, each of the P routers except $P_1$ has a subtending $PE$.) The P routers handle many types of SP traffic, e.g., VPN, Internet service, traditional wireline voice, or mobile voice and data traffic. There are typically many more PEs than P routers in the service provider network.

A VPN must satisfy three requirements: (*i*) Each customer has no access to information on any other VPN customers supported by the SP. (*ii*) Each customer must be allowed to use its own unicast or multicast address plan, regardless of the address plan used by other VPN customers supported by the SP. (*iii*) Each customer has no access to information about PE routers to which it is not attached, and no visibility into the SP backbone. In particular, each customer has no access to information about the P routers in the SP backbone and about the SP's internal addressing plan.

To meet these requirements, each PE stores a set of *Virtual Routing and Forwarding tables* (VRFs). A PE will have a VRF for a given VPN if there is a CE, belonging to that customer, which subtends the PE; in this case we say that the PE *is used* by the VPN. Thus, if there are $K$ PEs used by a VPN, each of the $K$ PEs will have a VRF for this VPN. A VRF for VPN $V$ has one forwarding table entry for each address prefix that is advertised by a CE for this customer to $V$. The information stored in a VRF on a given PE (call it $X$) for an address prefix $p$ includes the cost to reach the PE (call it $Y$) that is advertising $p$, and the interface out of $X$ on the best path to $Y$.

A VRF for a given VPN is populated as follows. Suppose $CE_p$ wants to advertise an address prefix $p$ throughout this VPN. Then $CE_p$ informs, typically using BGP, its local PE (call it $PE_p$) about $p$. So $PE_p$ installs $p$ in its VRF for this VPN. The service provider then distributes this address prefix, using BGP, to the other PEs used by this VPN. Once $p$ has been received by some remote PE (call it $PE_x$), $p$ is installed in the VRF for this

VPN on $PE_x$. Then $p$ is communicated, typically using BGP, to the CEs for this customer that subtend $PE_x$. If a CE for this customer subtending $PE_x$ wants to send a packet to destination $p$, then the CE forwards the packet to $PE_x$; the SP backbone network forwards it from $PE_x$ to $PE_p$; finally $PE_p$ forwards the packet to $CE_p$.

When the service provider forwards a customer packet from $PE_x$ to $PE_p$, the packet cannot be sent "natively" (i.e., unencapsulated) into the SP backbone network, since the destination address in the customer packet refers to a customer address, and the SP backbone routers have no information about customer addresses. Therefore, a customer's data packet is encapsulated to hide the customer destination address from the backbone routers. The encapsulation must carry the information needed by the SP backbone network to forward the packet from $PE_x$ to $PE_p$.

Since each customer independently chooses its own addressing plan, different customers might use the same address prefix. For example, two customers might both want to advertise the address prefix 10.1.2.0/24. BGP, used by the SP to distribute address prefixes within a VPN, needs a way of unambiguously determining which address prefix is associated with which VPN. This is accomplished by associating with each VRF a unique identifier, known as a *Route Distinguisher* (RD). Consider an address prefix $p$ installed in a VRF on some PE. Prepending the VRF RD to $p$ yields a *VPN-IP address*, which is customer specific. For example, if VRF 1 has $RD_1$ and VRF 2 has $RD_2$, then for the address prefix 10.1.2.0/24, the VPN-IP addresses $RD_1 : 10.1.2.0/24$ and $RD_2 : 10.1.2.0/24$ are now unique within the SP network.

Each VRF in a VPN is also associated with a set of *import Route Targets* (import RTs) and a set of *export Route Targets* (export RTs). When a PE uses BGP to advertise a particular VPN-IP address prefix from a particular VRF, it advertises the VRF's export RTs as *attributes* (i.e., associated parameters) of the VPN-IP address prefix. A VRF imports (i.e., accepts) a given VPN-IP address prefix if the address prefix carries (as an attribute) a RT that is one of the VRF's import RTs. Typically, for a given VPN all VPN-IP addresses prefixes are advertised with the same RT, and all VRFs for the VPN import that RT, so all VRFs learn all address prefixes for the VPN. Alternatively, RTs can provide fine grain control over which VRFs install which address prefixes. This is useful, e.g., in a hub and spoke topology in which spoke locations can talk only to hubs, but not to each other. RTs can also be used to control the distribution of multicast state, as discussed in Section 7.4 below.

The mechanisms for distributing customer address prefixes, RDs, and RTs among the PEs used by a VPN are called *control plane* procedures. The mechanisms for forwarding actual customer packets to the PEs used by a VPN are called *data plane* procedures. The service provider is responsible for implementing both the control plane and data plane procedures. In the above description of unicast VPNs, we have described the control plane procedures for distributing customer address prefixes. The data plane for unicast IP

VPNs is typically either native IP routing or MPLS forwarding, with tunnels between PEs to hide customer address prefixes from backbone routers.

Thus far we have only defined unicast VRFs. We define a *multicast VRF* (mVRF) as the routing/forwarding table associated with $(\star, g)$ or $(s, g)$ multicast state, for customer group $g$ (where $g$ is in the customer multicast address space) and customer source host $s$, for a given customer on a given PE. The entry in a given mVRF on a PE corresponding to a given $(\star, g)$ or $(s, g)$ state provides the RPF interface (towards the root of the customer shared or source tree) and the outbound interface list (OIL) over which incoming packets should be sent. Finally, a *multicast VPN* (mVPN) is a VPN which also contains a set of mVRFs with interfaces that can send multicast packets to each other. A multicast packet, sourced by customer host $s$ and for customer multicast group $g$, that is received by a PE is sent over the SP backbone network to one or more destination PEs, in order to reach the destination CEs for that customer which have subtending receiver hosts interested in $(\star, g)$ or $(s, g)$ packets. These destination PEs can be spread among the set of PEs used by the VPN.

We will discuss two current methods for implementing mVPNs: RFC 6037 ([93], see also [18], [83]) and a *BGP based* method ([3], [57], [58]). They differ in their control plane mechanisms for distributing customer multicast information: in RFC 6037 the PE-PE multicast control plane uses PIM, while in the BGP based method the PE-PE multicast control plane uses BGP. The BGP based method is a particular variant of the newer IETF standard mVPN scheme [91], which supports using either BGP or PIM for the PE-PE multicast control plane.

## 7.2 Draft-Rosen

For convenience we refer to RFC 6037 as *draft-Rosen*, since it was originally described in [92] and is well known by that name. The essence of the draft-Rosen method is that one multicast group is defined in the SP network for each multicast customer. The set of source and receiver nodes used by this group is precisely the set of PEs spanned by the customer mVPN. In the SP network, the multicast traffic is carried by a *default Multicast Distribution Tree* (default MDT), which is built using PIM-SM or SSM or BiDir. The default MDT is built in the SP network by running an instance of PIM, on the PE and P routers, that operates in the SP multicast address space (not in a customer multicast address space).

All multicast control traffic for this customer, as well as low bandwidth customer data traffic, is carried over the default MDT. For high bandwidth sources, a *data MDT*, typically created using SSM, can be used to send traffic only to PEs that have interested subtending receivers. We now describe default MDTs and data MDTs in more detail.

## 7.2.1 Default MDT

Consider a given mVPN $V$. An mVRF for $V$ is created on a PE by appropriate configuration of the PE, and this does not depend on having active multicast source hosts or receiver hosts for $V$ on any CE subtending the PE. If a PE has an mVRF for $V$, we say that the PE is *used* by $V$. Thus whether or not a given PE is used by $V$ is a static, rather than dynamic, condition. Let $\mathcal{PE}_V$ be the set of PEs used by $V$. The *Multicast Domain* for $V$ is the set of mVRFs for $V$ (thus the terms *multicast domain* and *multicast VPN* are synonymous).

The SP assigns to $V$ a unique multicast group address in the SP multicast address space; we denote this address by $G_V$. Previously we have exclusively used lower case $g$ to refer to a multicast address; here we use $G$ to emphasize that this is a multicast address in the service provider address space, and $G_V$ is the service provider multicast address assigned to $V$. We will continue to use lower case $g$ to refer to a multicast address in the customer multicast address space.

Regardless of the number of customer multicast groups supported in a given mVPN $V$, the SP creates for $V$, using PIM, one default MDT. The default MDT for $V$ connects all $|\mathcal{PE}_V|$ PEs in $V$ through the SP backbone network. No PE in $\mathcal{PE}_V$ is allowed to be pruned from the default MDT for $V$. The default MDT can be created using any PIM technique preferred by the service provider: PIM-SM, BiDir, or SSM. The default MDT for $V$ is created by running PIM on the PE and P routers in the SP network, using the address $G_V$. The address $G_V$ and the default MDT for the multicast customer $V$ are transparent to the customer. In order to use PIM-SM to create the default MDT for $V$, one or more RPs for the address $G_V$ must be deployed in the SP network. If there are multiple RPs for a given address $G_V$, they can be interconnected using MSDP.

A logical *multicast tunnel interface* (MTI) is defined on each mVRF for $V$. The MTI appears, to an mVRF, to be a LAN interface, and so each mVRF for $V$ views the other mVRFs for $V$ as belonging to that LAN. Thus a PIM adjacency can be established between each pair of mVRFs for $V$, even though each adjacency is actually formed between two distinct PEs separated by one or more P routers.

The default MDT for $V$ delivers both customer control data packets (e.g., customer multicast domain PIM joins and prunes) and customer multicast data packets for $V$ to each PE in $\mathcal{PE}_V$, whether or not there are interested receiver hosts behind these PEs. To distinguish between SP and customer packets, we write $cust(s, g)$ to refer to a *customer* source $s$ and a *customer* group $g$; i.e., an $(s, g)$ in the *customer* multicast domain. Referring to Figure 7.2, suppose a customer source host $s$ for mVPN $V$ behind $CE1$ sends to $CE1$ a customer multicast packet (*c-packet*, for brevity), addressed to customer multicast group $g$. Then $CE1$ forwards the c-packet to the mVRF for

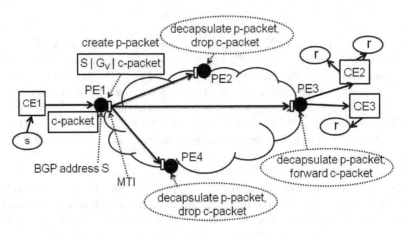

**Fig. 7.2** Packet forwarding in multicast VPN

$V$ on $PE1$. To send the c-packet over the MTI, the c-packet is encapsulated inside a service provider multicast packet (*p-packet*, for brevity). The source address $S$ of the p-packet is the address of $PE1$. The destination of the p-packet is the multicast address $G_V$ assigned by the SP. The p-packet travels over the default MDT for $V$ to all PEs (other than $PE1$) in $\mathcal{PE}_V$. At each such PE, the p-packet header is used to identify the mVRF for $V$ on that PE, and the c-packet is presented to that mVRF as having arrived over the associated MTI. If the $cust(s, g)$ or $cust(\star, g)$ OIL (in the customer multicast domain) is not empty, the c-packet is forwarded on the appropriate interfaces. Otherwise, if the $cust(s, g)$ or $cust(\star, g)$ OIL is empty, the c-packet is discarded. For example, in Figure 7.2, the p-packet is routed over the default tree for $V$ to PE2, PE3, and PE4. At PE2 and PE4, the c-packet is discarded, since there are no subtending receivers.

Only $(\star, G_V)$ and $(S, G_V)$ state is created on the backbone routers (i.e., the P-routers) for $V$. No customer multicast state is created on the backbone routers. This allows service providers to handle large numbers of multicast customers, each with potentially many multicast groups. However, the PEs will create state for both customer multicast groups and SP multicast groups.

This is illustrated in Figure 7.3, where the customer mVPN has ports on five CEs ($CE1$-$CE5$), and these CEs are connected to five PEs ($PE1$-$PE5$). There are five backbone P routers ($P1$-$P5$) and another backbone router serving as RP for the SP multicast groups. There is a source host for the

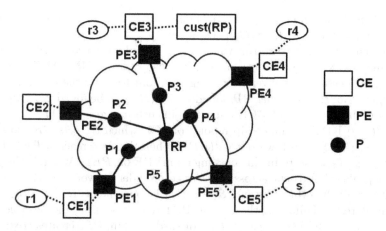

**Fig. 7.3** Default MDT

customer multicast group $g$ behind $CE5$, and receiver hosts for $g$ behind all the CEs except for $CE2$.

Assume the service provider runs PIM-SM. Then each PE in $\mathcal{PE}_V$ creates $(\star, G_V)$ state. With PIM-SM, each receiver node sends a join towards the RP (Section 3.7.2). With draft-Rosen, since we have assumed the SP runs PIM-SM, each PE in $\mathcal{PE}_V$ sends (upon initial configuration) a join towards $RP$, even if there are no interested customer receiver hosts (for any multicast group used by this customer) behind the PE. This creates $(\star, G_V)$ state on each node on the shortest path from each PE in $\mathcal{PE}_V$ to the RP. In Figure 7.3, since the shortest path from $PE1$ to $RP$ is via $P1$, then $(\star, G_V)$ state is created on $P1$ and $RP$. Similarly, $(\star, G_V)$ state is created on $P3$ by the $PE3$ join, and on $P4$ by the $PE4$ join. However, for $PE5$, its shortest path to $RP$ is via $P4$, so no $(\star, G_V)$ state is created on the backbone node $P5$.

With PIM-SM, a source node sends a unicast register message to the RP, and the RP in turns sends a join back towards the source node. With draft-Rosen, and since we have assumed the SP runs PIM-SM, each PE in $\mathcal{PE}_V$ registers with $RP$, even if there are no customer source hosts (for any multicast group used by this customer) behind the PE. This triggers a join from $RP$ towards the PE. If $S$ is the IP address of a given PE in $\mathcal{PE}_V$, then $RP$ sends an $(S, G_V)$ join towards the PE. Thus $(S, G_V)$ state is created on that PE and on each backbone router on the shortest path from $RP$ to that PE. In Figure 7.3, let $S_1$ be the IP address of $PE1$, and similarly for $S_2$, etc. The shortest path from $RP$ to $PE1$ is via $P1$, so $(S_1, G_V)$ state is created on $RP$, $P1$, and $PE1$. Similarly, $(S_2, G_V)$ state is created on $RP$, $P2$, and $PE2$; $(S_3, G_V)$ state is created on $RP$, $P3$, and $PE3$; and $(S_4, G_V)$ state is created on $RP$, $P4$, and $PE4$. The shortest path from $RP$ to $PE5$ goes via $P4$, so $(S_5, G_V)$ state is created on $RP$, $P4$, and $PE5$. Hence in general, if $|\mathcal{PE}_V| = K$, the maximum number of SP domain multicast states created on any PE in $\mathcal{PE}_V$, and on any SP backbone router, is at most $K+1$, namely, one $(\star, G_V)$ state and $K$ $(S, G_V)$ states.

The customer is free to run whatever PIM multicast protocol it chooses. Assume, for example, that the customer is also running PIM-SM. Then each interested customer receiver host triggers a join towards the customer RP. In Figure 7.3, the customer RP, labelled $cust(RP)$, is behind $PE3$. Suppose customer host $r_1$, behind $CE1$, wants to join customer group $g$. Then $r_1$ sends a $cust(\star, g)$ IGMP membership report to $CE1$, which triggers $CE1$ to send a $cust(\star, g)$ PIM join towards $cust(RP)$. When this join reaches $PE1$, $PE1$ creates $cust(\star, g)$ state in this customer's mVRF on $PE1$. At $PE1$, the join is encapsulated and sent across the service provider network (on the default MDT) to all the other PEs in $\mathcal{PE}_V$. Each PE, once it has received the join, decapsulates it. Following the normal PIM rules governing the reception of join messages, all the PEs except $PE3$ discard the join; $PE3$ creates $cust(\star, g)$ state on $PE3$ and forwards the join towards $cust(RP)$.

Since in our example we assumed the customer is running PIM-SM, each customer source host triggers a PIM register message towards the customer RP. In Figure 7.3, when the customer source host $s$ behind $CE5$ starts sending packets, $CE5$ sends, using unicast, a PIM register message towards $cust(RP)$. When the register message reaches $PE3$, it is forwarded to $cust(RP)$. The register message triggers $cust(RP)$ to send a $cust(s, g)$ join back towards $CE5$. The join is encapsulated at $PE3$, where it creates $cust(s, g)$ state, and is sent across the service provider network (on the default MDT, transparently to the customer) to $PE5$, where it is decapsulated and creates $cust(s, g)$ state, and is then sent to $CE5$.

## 7.2.2 Data MDTs

Assume the service provider is running draft-Rosen and is using PIM-SM to create the default MDTs, with a very high SPT-threshold (which prevents creation of shortest path trees, as described in Section 3.7.2). Then, as explained in Section 7.2.1 above, only two SP domain multicast states are created on each PE used by mVPN $V$, namely, $(\star, G_V)$ and $(S, G_V)$, where $S$ is the IP address of the PE. And with PIM-SM, as also explained above, for this $V$ at most $|\mathcal{PE}_V| + 1$ service provider domain multicast states are created on any P router. If the service provider uses BiDir for the default MDT, at most one $(\star, G_V)$ SP domain multicast state for $V$ is created on each P and PE router for $V$. However, with draft-Rosen, regardless of the multicast protocol run by the service provider, the default MDT sends all customer domain multicast packets for $V$ to each PE in $\mathcal{PE}_V$, whether or not there are interested receiver hosts behind the PE. If there are no interested hosts behind some PE, then bandwidth is wasted in the SP network delivering packets to this PE, and extra work is incurred at the PE to drop the unwanted packets. To address this concern, draft-Rosen allows creating a *data MDT* which will span only the receiver nodes of mVPN $V$. A data MDT is associated with a single $cust(s, g)$, and spans only those PEs in $\mathcal{PE}_V$ with subtending receiver hosts wishing to receive the $cust(s, g)$ stream.

Whereas the default MDT for $V$ is automatically established upon configuration of the mVPN $V$, a data MDT is established dynamically when the bandwidth of a $cust(s, g)$ multicast stream sent by some PE in $\mathcal{PE}_V$ exceeds a specified threshold. The recommended method [93] of creating a data MDT is to build a source tree using SSM. Consider a $cust(s, g)$ multicast stream for $V$ sent by a customer source host $s$ subtending a CE, which in turn subtends some PE in $\mathcal{PE}_V$. Let $S$ be the IP address of the PE. When the bandwidth of the $cust(s, g)$ stream exceeds a specified threshold, the PE selects an address $\bar{G}$ from its configured SSM range (in the SP domain multicast address space) and advertises $(S, \bar{G})$, over the default MDT, to all the PEs in $\mathcal{PE}_V$. A PE wishing to receive the $cust(s, g)$ stream may then send a $(S, \bar{G})$ join towards $S$ to receive the stream. To give receiver PEs time to join the data MDT, the source node waits some time before switching from the default MDT to the data MDT. Once the source node sends the $cust(s, g)$ stream on the data MDT, it stops sending the stream over the default MDT. Low bandwidth customer streams, as well as all customer and SP control packets, continue to flow over the default MDT, so they will reach all PEs in $\mathcal{PE}_V$.

The data MDT persists as long as the $cust(s, g)$ stream bandwidth exceeds the threshold. If the bandwidth falls below the threshold, the stream will again be sent over the default MDT. The $(S, \bar{G})$ can be cached in each PE in $\mathcal{PE}_V$ for use if the $cust(s, g)$ bandwidth again exceeds the threshold, although this cache may age out after some interval, if not refreshed.

## 7.3 A General Approach to Multicast VPNs

A more general approach to building multicast VPNs, described in 2009 by Rosen and Aggarwal [91], allows a wide range of methods for building trees in the service provider network. We require some terminology. *Auto-discovery* (AD) is the process employed by one PE used by an mVPN to learn about other PEs used by the mVPN. Two methods for AD are discussed in [91]. One way has each PE advertise, using BGP, its membership in an mVPN. The other method, defined in draft-Rosen, requires the creation of a PIM-SM or BiDir shared tree, over which PEs auto-discover each other simply by joining the shared tree and sending PIM hello messages over the tree.

A *Provider Multicast Service Instance* (PMSI) is a service, created in the service provider network and provided by means of one or more tunnels, used for forwarding customer domain mVPN multicast traffic to some or all of the

PEs used by the mVPN. The term PMSI is generic, and does not specify how the tunnel is created. Seven types of tunnels that may be used to create a PMSI are: PIM-SM trees, BiDir trees, SSM trees, P2MP LSPs created by either RSVP-TE (see [73]) or mLDP, MP2MP LSPs created by mLDP, and unicast tunnels used with replication (Section 1.3).

PMSIs come in two main flavors, inclusive and selective. An *inclusive* PMSI (I-PMSI) is a PMSI which connects to each PE used by the mVPN. We can further divide I-PMSIs into *multidirectional inclusive* PMSIs (MI-PMSIs), which allow any PE used by the mVPN to send packets which will be received by all other PEs used by the mVPN, and *unidirectional inclusive* PMSIs (UI-PMSIs), which allow only a particular PE used by the mVPN to send packets which will be received by all other PEs used by the mVPN. An MI-PMSI can be instantiated, e.g., by a Bidir shared tree, by an MP2MP LSP spanning every PE used by the mVPN, by unicast replication, and by the draft-Rosen default MDT. A UI-PMSI can be instantiated, e.g., by a P2MP LSP spanning every PE used by the mVPN.

A *selective* PMSI (S-PMSI) is a PMSI which spans only a subset of the PEs used by the mVPN. An S-PMSI can be instantiated by the draft-Rosen data MDT using SSM. An S-PMSI can also be instantiated by a P2MP LSP spanning only a subset of the PEs used by the mVPN. An S-PMSI is desirable for high-bandwidth $cust(s, g)$ flows which should not be sent to a PE unless there is an interested subtending receiver host.

There must be at least one PMSI for each mVPN, and an mVPN might also use multiple S-PMSIs, each carrying a different set of $cust(s, g)$ streams for that mVPN. It is also possible to use aggregate tunnels, which carry traffic for multiple mVPNs (Chapter 6).

## 7.4 BGP mVPNs

An alternative to the draft-Rosen method for building mVPNs is is to let BGP advertise which PEs have interest in which $cust(s, g)$ or $cust(\star, g)$ streams (recall that $cust(s, g)$ refers to the $(s, g)$ in the *customer* multicast domain, as opposed to the SP multicast domain). With this approach, a PE will join a PMSI only if it has subtending hosts interested in receiving the $cust(s, g)$ or $cust(\star, g)$ streams. The motivation for this approach is that, for unicast VPNs, BGP is currently the method of choice for distributing unicast customer address prefixes among the PEs used by a VPN, and BGP can be extended to also carry customer domain multicast information. An mVPN built this way is called a *BGP mVPN* ([3], [57], [58]).

The BGP mVPN approach defines seven types of control messages, used to carry auto-discovery (AD) messages and to set up a PMSI. (In [57], [73], and IETF documents, these control message are called *routes*; since this use of *route* is entirely different than using *route* to mean *path*, to avoid confusion we call them *message types*.) Message types 1-4 below provide all the control mechanisms to set up an I-PMSI or S-PMSI for an mVPN.

- An *Intra-AS I-PMSI AD message* (type 1) is originated by each PE, in a given Autonomous System (AS), belonging to a given mVPN. This message informs other PEs of the originating PE's ability to participate in the mVPN, and can also be used to specify an I-PMSI to be used as the default method of carrying multicast traffic for that mVPN.
- For brevity, we ignore message type 2, which is associated with inter-AS multicast.
- An *S-PMSI AD message* (type 3) is originated by a source PE to announce that the source PE will be using an S-PMSI to transmit a given $cust(s, g)$ stream, and to specify the method used to instantiate that S-PMSI.
- A *leaf AD message* (type 4) is originated by a PE, in response to receiving a type 3 message, if the PE has subtending hosts interested in receiving the stream transmitted on the S-PMSI; this message type is used by the source PE to discover the leaf nodes of an S-PMSI rooted at the source PE. A type 4 message is not needed if the S-PMSI is instantiated by a receiver-initiated multicast protocol, such as PIM or mLDP, since in those cases the source node does not need to discover the leaf nodes.
- A *Source Active (SA) AD message* (type 5) is originated by a PE in response to a customer source (on a host subtending the PE) becoming active; this message is used by other PEs to learn the identity of active multicast sources. Type 5 messages are used only with PIM-SM, to create shortest path trees. They provide a way to implement RP-bit Prune functionality (Section 3.7.2), needed when a PE wants to join the $cust(\star, g)$ shared tree, but not for certain customer sources.

The next two message types, 6 and 7, carry customer domain joins from a receiver PE to a source PE. In BGP mVPNs these two message types are carried using BGP, rather than using PIM as with draft-Rosen; see [3] more details.

- A *shared tree join message* (type 6) is originated by a receiver PE when it receives a shared tree $cust(\star, g)$ join from a subtending CE.
- A *source tree join message* (type 7) is originated by a receiver PE when it receives a $cust(s, g)$ join from a subtending CE.

In Section 7.1 we introduced route targets (RTs) for unicast VPNs. These RTs are used to ensure that each unicast address prefix in a VPN is made known to all other PEs used by the VPN. A second type of route target, which we call a *c-mVRF RT* (this is a shorter and more descriptive name than the name *VRF route import extended community RT* used in [57]), is required by BGP mVPNs. Each PE creates a unique c-mVRF RT for each VPN that uses the PE. Suppose PE $X$ wants to join some $cust(s, g)$ or $cust(\star, g)$ tree, and $X$ determines that the customer source or customer RP for this $g$ subtends

PE $Y$. Then the c-mVRF RT associated with $Y$ is attached to the type 6 or type 7 message that $X$ sends; if the message is received by a PE other than $Y$, then that PE can discard the message.

Assuming the service provider is building S-PMSIs and instantiating them as P2MP trees spanning only the source and the receiver nodes, we now use Figure 7.4 to explain how customer domain joins are forwarded in BGP mVPNs. Consider a given mVPN $V$ and suppose PEs $R$ and $S$ are two

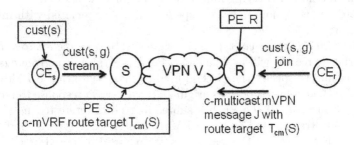

**Fig. 7.4** BGP mVPN

PEs used by the mVPN. Suppose $CE_s$ subtends $S$ and $CE_r$ subtends $R$. Let $T_{cm}(S)$ denote the c-mVRF route target for mVPN $V$ on $S$. The BGP protocol ensures that $R$ learns of $S$ and $T_{cm}(S)$.

Suppose a customer domain source host $s$ subtending $CE_s$ is sending to multicast group $g$, and a customer host subtending $CE_r$ wishes to receive the $cust(s,g)$ stream. So $CE_r$ sends a $cust(s,g)$ join to $R$, which ultimately causes $S$ to send a $cust(s,g)$ join to $CE_s$. This is accomplished as follows.

1. $R$ receives a $cust(s,g)$ join from $CE_r$.
2. $R$ finds $S$ and $T_{cm}(S)$ in its unicast VRF for VPN $V$
3. $R$ creates a type 7 source tree join message (call it $J$) and attaches $T_{cm}(S)$ to $J$
4. $R$ sends, using BGP, the join $J$ to $S$
5. $S$ accepts $J$, since $T_{cm}(S)$, the c-mVRF route target for $V$ at $S$, matches the c-mVRF route target attached to $J$.
6. $S$ translates $J$ into a join in the customer domain.
7. $CE_s$ processes the $cust(s,g)$ join from $CE_r$ using the customer domain instance of PIM.

In BGP mVPNs, BGP is used not just for transmitting customer domain joins and prunes, but also for advertising the tunnel type to the PEs in $\mathcal{PE}_V$. Once the tunnel type is signalled by the source PE to the other PEs in $\mathcal{PE}_V$, the source and receiver PEs *bind* the mVRF for $V$ to the locally configured tunnel (analogous to the use of the MTI in draft-Rosen).

To reduce the amount of SP domain multicast state on P routers, the various aggregation techniques reviewed in Chapter 6 can be utilized with BGP mVPNs [69]. For example, the *aggregate selective tree* approach is based upon the observation that, if some mVPN $V$ requires traffic to be delivered to a set $\mathcal{PE}_V$ of PEs, then it is possible that there are other mVPNs that require traffic to be delivered to exactly the same set. Therefore, with each possible set $\mathcal{S}$ of PEs we can associate a given multicast tree $\mathcal{T}(\mathcal{S})$. The advantage of this approach is that traffic is delivered only to those PEs behind which are receivers (hence the adjective *selective*), so there is no leakage. However, multicast group membership is dynamic, so the set of receiver PEs will in general change even during the duration of a multicast stream. Another drawback of this method is that, if there are $N$ PEs, then there are $2^N$ possible subsets, and hence that many possible trees. Alternatively, the *aggregate inclusive tree* approach uses a smaller number of possible trees, but allows leakage.

## 7.5 Monitoring Multicast VPN Performance

Monitoring IP multicast has traditionally been difficult. The difficulty is compounded with multicast VPNs, since multicast exists in both the customer and service provider (SP) domains. The VMScope tool of Breslau, Chase, Duffield, Fenner, Mao, and Sen [17] measures packet loss rates and delays by using tunnels to send and receive multicast traffic across the SP backbone, where it appears like, and is treated the same as, customer multicast traffic carried by the SP. The tool, which runs on a single monitoring host without any special hardware, creates a tunnel, using GRE encapsulation (Section 1.4), with each PE router being monitored. In the following discussion of VMScope, $G$ refers to a multicast group address in the SP multicast address space, and not to a customer domain multicast group.

Consider a tunnel connecting the monitoring host $h$ with a remote PE $n$. First, $h$ sends an IGMP membership report for $G$ across the tunnel to $n$. Then $n$ joins the multicast tree for $G$, using whatever multicast routing protocol has been deployed. When $n$ receives a multicast packet addressed to $G$, it forwards (using unicast) the packet across the tunnel to $h$, as if $h$ were a directly attached host. The via routers between $n$ and $h$ need not support multicast. Besides receiving packets, $h$ can also source multicast packets, and send them (using unicast) to $n$ across the tunnel. When $n$ receives the packet, it removes the GRE header and forwards the packets as dictated by the multicast forwarding table for $G$.

Now suppose there are $K$ PEs, $n_1, n_2, \cdots, n_K$, and a tunnel $T_k$ from $h$ to each of the $K$ PEs. VMScope computes the round trip delay (using the *ping* utility) and the packet loss between $h$ and each $n_k$ , and assigns half this delay and packet loss to each direction of the tunnel. For group $G$, host $h$ sends multicast packets to $n_1$ across $T_1$, and sends, for $2 \leq k \leq K$, an IGMP membership report to $n_k$ across $T_k$. PE $n_1$ decapsulates packets received from $h$, and forwards them as normal multicast packets to the tree for $G$. For $2 \leq k \leq K$, since $n_k$ joined the tree for $G$, these packets will be forwarded to $n_k$, which then encapsulates them and sends them over $T_k$ to the monitoring host $h$. Thus, for each packet $h$ sends to $n_1$, it receives back $K - 1$ copies of the packet, assuming no packet loss.

Each packet originated by $h$ includes an application layer header which contains a sequence number. Also, $h$ stores the time it sends each packet. For each packet received by $h$, the PE sending it is known, since if the packet is received over tunnel $T_k$ it came from $n_k$. Thus $h$ can compute both the loss rate and the total round trip delay from $h$ to $n_1$ to each $n_k$ and back to $h$. Since the delay on each tunnel has been computed, subtracting out the tunnel delays yields the delay from $n_1$ to each $n_k$.

We mention also the work of Duffield, Horowitz, Lo Presti, and Towsley [36], who discuss methods to infer the topology of a multicast tree from end-to-end measurements taken when probe packets are sent from the source to the set of receivers.

## 7.6 Empirical Study of Enterprise Multicast Traffic

Studying production enterprise customer multicast traffic can help determine if the current multicast protocols are efficient, or if new protocols are desirable, and also can help in network planning and network management. In the first study of enterprise multicast traffic, Karpilovsky, Breslau, Gerber, and Sen [61] provides statistics on the $cust(\star, g)$ and $cust(s, g)$ flows (i.e., streams) carried by a service provider running draft-Rosen.

The authors collected data for 35 days, polling each PE every 5 minutes, and amassing about 88 million records. In addition, the industry (e.g., health services, manufacturing, or finance) for each mVPN was available. Only limited analysis was performed on the default MDT, since this tree also carries PIM hello messages (Section 3.7), which make it difficult to tell when customer domain data flows start or stop, and since higher bandwidth customer flows, which are of greater interest, are carried on data MDTs. Nearly all (99.5%) of the default MDTs have bandwidth less than 5 Kbps, but a very small percentage has flows above 100 Kbps; this may occur if the flow duration is not sufficiently long to trigger a switchover to a data MDT, or if a high bandwidth flow lasted only a short period of time on the default MDT before switching to the data MDT. Thus the remainder of the discussion of these results concerns the data MDTs only.

In order to identify sets of behavior, each $cust(s, g)$ flow was assigned to one of 4 buckets; these buckets are identified in the first column of Table 7.1 below. (The study actually considered from 1 to 25 buckets, and picked 4 buckets as

| bucket | duration | BW | peak BW | max recv's | avg recv's | % of flows |
|--------|----------|-----|---------|------------|------------|------------|
| unicast | 29 hrs, 6 min | 12 Mbps | 23 Mbps | 1.4 | 1.2 | 0.1 |
| limited recv's | 1 hour, 12 min | 40 Kbps | 56 Kbps | 2.7 | 2.1 | 86.5 |
| long lived | 28 days | 605 Kbps | 984 Kbps | 9.5 | 3.1 | 0.05 |
| well fitted | 59 minutes | 21 Kbps | 31 Kbps | 25.4 | 19.6 | 13.3 |

**Table 7.1** The four cluster centers

the value for which a significant fraction of the variance was explained, while keeping the number of buckets reasonably small.) In the header row, "BW" is shorthand for "bandwidth", and "recv's" is shorthand for "receivers". For the *unicast* bucket, flows are long lived, with high throughput, and very few receiver PEs. Surprisingly, almost 50% of the data MDT flows have only a single receiver PE; it is desirable to identify such flows and send them across the service provider backbone using unicast encapsulation, which incurs less routing state overhead than multicast. However, about 20% of data MDT flows reach at least 10 different PEs during the flow lifetime; multicast works well for such flows. The *well fitted* bucket is so named since it exhibits a large number of receivers, moderate length flows, and moderate bandwidth flows. For each industry, the percentage of data MDT flows in each bucket is given in Table 7.2.

| category | % unicast | % limited recv's | % long lived | % well fitted |
|----------|-----------|------------------|--------------|---------------|
| health services | 0 | 94.3 | 2.9 | 2.9 |
| manufacturers | 0.3 | 78.5 | 0.03 | 21.1 |
| retailers | 0 | 94 | 0 | 6 |
| finance | 0.4 | 86.1 | 0.02 | 13.5 |
| tech | 0 | 99.6 | 0 | 0.4 |
| information services | 0 | 75.1 | 0.04 | 24.8 |
| natural resources | 0 | 100 | 0 | 0 |
| other | 0.3 | 98.5 | 0 | 1.2 |
| average | 0.1 | 86.5 | 0.05 | 13.3 |

**Table 7.2** The percentage of flows in each bucket

The *unicast* flows were almost entirely limited to the manufacturing and financial industries. However, even for these industries, the percentage of flows in the unicast bucket is quite small (0.3% and 0.4%). The *limited receivers* bucket gets the lion's share of flows in all industries (86.5% on average) suggesting that many applications exhibit this behavior, and that improvements in general multicast protocols should focus on this bucket. By far the highest percentage of flows in the *long lived* bucket occurred for the health services industry (2.9%). Since this bucket has fairly high bandwidth and not a small number of receivers, some multicast optimization for this category may be possible. Finally, for the *well fitted* bucket, the percentage of flows varied greatly across industries, so it is difficult to characterize the applications comprising this bucket, and thus difficult to see how multicast might be optimized.

# References

1. Adjih C, Georgiadis L, Jacquet P, Szpankowski W (2001) Is the Internet Fractal? The Multicast Power Law Revisited. INRIA Research Report 4157, Hipercom Project, ISSN 0249-6399
2. Adjih C, Georgiadis L, Jacquet P, Szpankowski W (2006) Multicast Tree Structure and the Power Law. IEEE Trans. on Information Theory 52:1508-1521
3. Aggarwal R, Rosen E, Morin T, Rekhter Y (2009) BGP Encodings and Procedures for Multicast in MPLS/BGP IP VPNs. IETF draft-ietf-l3vpn-2547bis-mcast-bgp-08
4. Ahlswede R, Cai N, Li SYR, Yeung RW (2000) Network Information Flow. IEEE Trans. on Information Theory 46:1204-1216
5. Ahuja RK, Magnanti TL, Orlin JB (1993) Network Flows. Prentice-Hall, New Jersey
6. Almeroth K (2000) The Evolution of Multicast: From the MBone to Interdomain Multicast to Internet2 Deployment. IEEE Network Jan/Feb:10-20
7. Arya V, Turletti T, Kalyanaraman S (2005) Encodings of Multicast Trees. In: Networking 2005:922-1004 Lecture Notes in Computer Science. Springer, New York, NY
8. ATM Forum Technical Committee (2002) Private Network-Network Interface Specification Version 1.1 af-pnni-0055.002
9. Ballardie T, Cain B, Zhang Z (1998) Core Based Trees (CBT version 3) Multicast Routing. IETF draft-ietf-idmr-cbt-spec-v3-00.txt
10. Ballardie T, Francis P, Crowcroft J (1993) Core Based Trees: An Architecture for Scalable Inter-Domain Multicast Routing. ACM SIGCOMM'93 and Computer Communication Review 23:8595
11. Bates T, Chandra R, Katz D, Rekhter Y (1998) Multiprotocol Extension for BGP-4. IETF draft-ietf-idr-rfc4760bis-03.txt
12. Bejerano Y, Busi I, Ciavaglia L, Hernandez-Valencia E, Koppol P, Sestito V, Vigoureux M (2009) Resilent Multipoint Networks Based on Redundant Trees. Bell Labs Technical Journal 14:113-130
13. Ben Ali N, Belghith A, Moulierac J, Molnar M (2008) QoS Multicast Aggregation Under Multiple Additive Constraints. Computer Communication 31:3654-3578
14. Bertsekas D, Gallager R (1992) Data Networks 2nd ed. Prentice Hall, New Jersey
15. Bhattacharyya S, ed. (2003) An Overview of Source-Specific Multicast (SSM). IETF RFC 3569, July 2003
16. Blaszczyszyn B, Tchoumatchenko K (2004) Performance Characteristics of Multicast Flows on Random Trees. Stochastic Models 20:341-361
17. Breslau L, Chase C, Duffield N, Fenner B, Mao Y, Sen S (2006) VMScope - A Virtual Multicast VPN Performance Monitor. SIGCOMM '06 Workshops in Pisa, Italy
18. Cai, Y (2004) Deploying IP Multicast VPN. Cisco Networkers 2004 Conference www.cisco.com/networkers/nw04/presos/docs/RST-2702.pdf
19. Chaintreau A, Baccelli F, Diot C (2001) Impact of Network Delay Variations on Multicast Sessions with TCP-like Congestion Control. In: Proc. IEEE INFOCOM 2001 in Anchorage, AK 1133-1142
20. Chalmers RC, Almeroth KC (2003) On the Topology of Multicast Trees. IEEE/ACM Trans. on Networking 11:153-165
21. Chen X, Wu J (2003) Multicasting Techniques in Mobile Ad Hoc Networks. In: Ilyas M, Dorf RC (eds) The Handbook of Ad Hoc Wireless Networks. CRC Press, Boca Raton, FL
22. Chuang J, Sirbu M (1998) Pricing Multicast Communication: a Cost Based Approach. Proc. INET '98 in Geneva, Switzerland. http://repository.cmu.edu/tepper/454

23. Cisco Systems (2002) IP Multicast Technology Overview White Paper: Version 3 http://www.cisco.com/en/US/docs/ios/solutions_docs/ip_multicast/White_papers/mcst_ovr.pdf

24. Cisco Systems (2004) Guidelines for Enterprise IP Multicast Address Allocation. http://www.cisco.com/warp/public/cc/techno/tity/prodlit/ipmlt_wp.pdf

25. Cisco Systems (2008) Cisco IOS IP Multicast Configuration Guide Release 12.2SX. Cisco Press, Indianapolis, IN

26. Ciurea IM (2005) State-Efficient Point to Multi-Point Trees using MPLS Multicast. M.S. Thesis, Simon Fraser University (condensed version at http://doi.ieeecomputersociety.org/10.1109/ISCC.2005.132)

27. Dalal, YK (1977) Broadcast Protocols in Packet Switched Computer Networks. Ph.D. dissertation, Stanford University, DSL Technical Report 128

28. Dalal YK, Metcalfe RM (1978) Reverse Path Forwarding of Broadcast Packets. Communications of the ACM 21:1040-1048

29. Davie B, Doolan P, Rekhter Y (1998) Switching in IP Networks. Morgan Kaufman, San Francisco

30. Davie B, Rekhter Y (2000) MPLS: Technology and Applications. Morgan Kaufman, San Francisco

31. Deering S (1989) Host Extensions for IP Multicasting. IETF Network Working Group Request for Comments 1112

32. Deering S (1991) Multicast Routing in a Datagram Internetwork. Ph.D. dissertation, Stanford University

33. Deering S, Estrin D, Farinacci D, Jacobson V, Liu CG, Wei L (1994) An Architecture for Wide-Area Multicast Routing. Proc. SIGCOMM '94 in London, England 126-135

34. Deering S, Estrin D, Farinacci D, Jacobson V, Helmy A, Meyer D, Wei L (1998) Protocol Independent Multicast Version 2 Dense Mode Specification. IETF draft-ietf-pim-v2-dm-01.txt

35. Diot C, Dabbous W, Crowcroft J (1997) Multipoint Communication: A Survey of Protocols, Functions, and Mechanisms. IEEE Journal on Selected Areas in Communications 15:277-290

36. Duffield NG, Horowitz J, Lo Presti F, Towsley D (2002) Multicast Topology Inference from Measured End-to-End Loss. IEEE Trans. on Information Theory 48:26-45

37. Estrin D, Farinacci D, Helmy A, Thaler D, Deering S, Handley M, Jacobsen V, Liu C, Sharma P, Wei L (1997) Protocol Independent Multicast-Sparse Mode (PIM-SM): Protocol Specification. IETF RFC 2117

38. Estrin D, Farinacci D, Helmy A, Thaler D, Deering S, Handley M, Jacobsen V, Liu C, Sharma P, Wei L (1998) Protocol Independent Multicast-Sparse Mode (PIM-SM): Protocol Specification. IETF RFC 2362

39. Farinacci D, Li T, Hanks S, Meyer D, Traina P (2000) Generic Routing Encapsulation (GRE). IETF RFC 2784

40. Fei A, Cui J, Gerla M, Faloustsos M (2001) Aggregated Multicast: an Approach to Reduce Multicast State. Proc. IEEE Global Telecommunications Conference (GLOBECOM '01) in San Antonio Texas 1595-1599

41. Fenner B, Handley M, Holbrook H, Kouvelas I (2006) Protocol Independent Multicast - Sparse Mode (PIM-SM): Protocol Specification (Revised). IETF RFC 4601

42. Garey MR, Johnson DS (1979) Computers and Intractability: A Guide to the Theory of NP-Completeness. W.H. Freeman, San Francisco

43. Guichard J, Pepelnjak I (2001) MPLS and VPN Architectures. Cisco Press, Indianapolis, IN

44. Haberman B (2002) Allocation Guidelines for IPv6 Multicast Addresses. IETF RFC 3307

45. Hać A, Zhou K (1999) A New Heuristic Algorithm for Finding Minimum-Cost Multicast Trees with Bounded Path Delay. Int. J. of Network Management 9:265-278

46. Handley M, Kouvelas I, Speakman T, Vicisano V (2007) Bi-directional Protocol Independent Multicast (BIDIR-PIM). http://tools.ietf.org/html/draft-ietf-pim-bidir-09

47. Hinden R, Deering S (2006) IP Version 6 Addressing Architecture. IETF RFC 4291

48. Holbrook H, Cain B (2006) Source-Specific Multicast for IP. IETF RFC 4607

49. https://datatracker.ietf.org/wg/l2vpn/

50. http://www.iana.org

51. Huang TL, Lee DT (2005) Comments and An Improvement on "A Distributed Algorithm of Delay-Bounded Multicast Routing for Multimedia Applications in Wide Area Networks". IEEE/ACM Trans. on Networking 13:1410-1411

52. Huitema C (1995) Routing in the Internet. Prentice Hall, New Jersey

53. Huntting B, Mertz D (2003) Multicasting Strategies: Protocols and Topologies. http://gnosis.cx/publish/programming/multicast_2.txt

54. Jia X (1998) A Distributed Algorithm of Delay-Bounded Multicast Routing for Multimedia Applications in Wide Area Networks. IEEE/ACM Trans. on Networking 6:828-837

55. Jiang Y, Wu M, Shu W (2004) A hierarchical overlay multicast network. In Proc. 2004 IEEE International Conference on Multimedia and Expo (ICME '04) 1047-1050

56. Junhai L, Danxia Y, Liu X, Mingyu F (2009) A Survey of Multicast Routing Protocols for Mobile Ad-Hoc Networks. IEEE Communications Surveys & Tutorials 11:First Quarter

57. Juniper Networks (2009) Understanding Junos Next-Generation Multicast VPNs. www.juniper.net/us/en/local/pdf/whitepapers/2000320-en.pdf

58. Juniper Networks (2010) Emerging Multicast VPN Applications. www.juniper.net/us/en/local/pdf/whitepapers/2000291-en.pdf

59. Juniper Networks (2010) MSDP Overview. www.juniper.net/techpubs/software/junos/junos94/swconfig-multicast/msdp-overview.html

60. Juniper Networks (2011) Unlocking Television over the Internet: Introduction to Automatic IP Multicast without Explicit Tunnels. White paper 2000375-001-EN

61. Karpilovsky E, Breslau L, Gerber A, Sen S (2009) Multicast Redux: A First Look at Enterprise Multicast Traffic. In Proc. WREN '09 in Barcelona, Spain 55-63

62. Kernen T, Simlo S (2010) AMT - Automatic IP Multicast without Explicit Tunnels. EBU Technical Review. European Broadcasting Union, Gevena, Switzerland Q4

63. Kou LT, Makki K (1987) An Even Faster Approximation Algorithm for the Steiner Tree Problem in Graphs. Congressus Numerantium 59:147-154

64. Kou L, Markowsky G, Berman L (1981) A Fast Algorithm for Steiner Trees. Acta Informatica 15:141-145

65. Lao L, Cui JH, Gerla M, Maggiorini D (2005) A comparative study of multicast protocols: top, bottom, or in the middle?. In Proc. IEEE INFOCOM 2005 4:2809-2814

66. Li J, Mirkovic J, Ehrenkranz T, Wang M, Reiher P, Zhang L (2008) Learning the Valid Incoming Direction of IP Packets. Computer Networks 52:399-417

67. Lua EK, Zhou X, Crowcroft J, Van Mieghem P (2008) Scalable Multicasting with Network-Aware Geometric Overlay. Computer Communications 31:464-488

68. Malkin GS, Steenstrup ME (1995) Distance-Vector Routing. In: Steenstrup M (ed) Routing in Communications Networks. Prentice Hall, Englewood Cliffs, New Jersey

69. Martinez-Yelmo I, Larrabeiti D, Soto I, Pacyna P (2007) Multicast Traffic Aggregation in MPLS-Based VPN Networks. IEEE Communications Magazine, October:78-85

70. Meyer D, Lothberg P (2001) GLOP Addressing in 233/8. IETF RFC 3180

71. Microsoft Research (2011) Avalanche: File Swarming with Network Coding. http://research.microsoft.com/en-us/projects/avalanche

72. Minei I, Kompella K, Wijnands I, Thomas B (2010) Label Distribution Protocol Extensions for Point-to-Multipoint and Multipoint-to-Multipoint Label Switched Paths. IETF draft-ietf-mpls-ldp-p2mp-09

73. Minei I, Lucek J (2011) MPLS-Enabled Applications, 3rd edn. John Wiley & Sons, West Sussex, United Kingdom

74. Moulierac J, Guitton A, Molnár M (2006) Hierarchical Aggregation of Multicast Trees in Large Domains. J. of Communications 1:33-44

75. Moy JT (1994) Multicast Extensions to OSPF. IETF RFC 1584

76. Moy JT (1994) MOSPF: Analysis and Experience. IETF RFC 1585

77. Moy JT (1998) OSPF: Anatomy of an Internet Routing Protocol. Addison-Wesley, Reading, MA

78. Napierala N, Rosen EC, Wijnands I (2011) A Simple Method for Segmenting Multicast Tunnels for Multicast VPNs. IETF draft-rosen-l3vpn-mvpn-segments-00.txt

79. Napierala M, Rosen EC, Wijnands I (2011) Using LDP Multipoint Extensions on Targeted LDP Sessions. IETF draft-napierala-mpls-targeted-mldp-00.txt

80. Novak R, Rugelj J, Kandus G (2001) Steiner Tree Based Distributed Multicast Routing in Networks. In: Cheng X, Du D (eds) Steiner Trees in Industry. Kluwer Academic, Norwell, MA 327-351

81. Oliveira CAS, Pardalos PM, Resende MGC (2006) Optimization Problems in Multicast Tree Construction. In: Resende MCG, Pardalos PM (eds) Handbook of Optimization in Telecommunications. Springer, New York, NY 701-731

82. Pai V, Kumar K, Kamilmani K, Sambamurthy V, Mohr AE (2005) Chainsaw: Eliminating Trees from Overlay Multicast. In: Peer-to-Peer Systems IV. Springer, New York, NY 127-140

83. Pepelnjak I, Guichard J, Apcar J (2003) MPLS and VPN Architectures, Volume II. Cisco Press, Indianapolis, IN

84. Perlman, R (2000) Interconnections, 2nd edn. Addison-Wesley, Boston, MA

85. Phillips G, Shenker S, Tangmunarunkit H (1999) Scaling of Multicast Trees: Comments on the Chuang-Sirbu Scaling Law. In: Proc. ACM SIG-COMM '99 in Cambridge, MA

86. Radoslavov P, Estrin D, Govindan R, Handley M, Kumar S, Thaler D (2000) The Multicast Address-Set Claim (MASC) Protocol. IETF RFC 2909

87. Ratnasamy S, Francis P, Handley M, Karp R, Shenker S (2001) A Scalable Content-Addressable Network. In: Proc. SIGCOMM'01 in San Diego, CA 161-172

88. Ratnasamy S, Handley M, Karp R, Shenker S (2001) Application-level Multicast using Content-Addressable Networks. In: Networked Group Communication, Lecture Notes in Computer Science 2233. Springer, New York, NY

89. Ritvanen K (2004) Multicast Routing and Addressing. Seminar on Interworking, Spring 2004, Helsinki University of Technology, www.tml.tkk.fi/Studies/T-110.551/2004/papers/Ritvanen.pdf

90. Rosen, EC (2010) private communication

91. Rosen EC, Aggarwal R (2010) Multicast in MPLS/BGP IP VPNs. IETF draft-ietf-l3vpn-2547bis-mcast-10.txt

92. Rosen EC, Cai Y, Wijnands I (2004) Multicast in MPLS/BGP IP VPNs. http://tools.ietf.org/html/draft-rosen-vpn-mcast-07

93. Rosen EC, Cai Y, Wijnands I (2010) Cisco Systems' Solution for Multicast in BGP/MPLS IP VPNs. http://tools.ietf.org/html/rfc6037

94. Rosenberg E (1987) A New Iterative Supply/Demand Router with Rip-Up Capability for Printed Wire Boards. In: Proc. 24 Design Automation Conference 721-726

95. Rosenberg E (2005) Hierarchical PNNI Addressing by Recursive Partitioning. In: Proc. IEEE Pacific Rim Conference on Communications, Computers, and Signal Processing (PacRim 2005) 133-136

96. Rosenberg E (2005) Hierarchical Topological Network Design. IEEE/ACM Trans. on Networking 13:1402-1409

97. Secci S, Rougier JL, Pattavina A (2007) Constrained Steiner Problem with Directional Metrics. In: Pereira P (ed) EuroFGI Workshop on IP QoS and Traffic Control in Lisbon, Portugal

98. Semeria C, Maufer T (1996) Introduction to IP Multicast Routing. IETF draft-rfced-info-semeria-00.txt

99. Shen CC, Li K, Jaikaeo C, Sridhara V (2008) Ant-Based Distributed Constrained Steiner Tree Algorithm for Jointly Conserving Energy and Bounding Delay in Ad Hoc Multicast Routing. ACM Trans. on Autonomous and Adaptive Systems 35, No. 1, Article 3

100. Sprintson A (2010) Network Coding and its Applications in Communication Networks. In: Cormode G, Thottan M (eds) Algorithms for Next Generation Networks. Springer, New York, NY

101. Stallings WJ (1998) Data and Computer Communications 2nd edn. Macmillan, New York

102. Stewart JW (1999) BGP4: Inter-Domain Routing in the Internet. Addison-Wesley, Reading, MA

103. Surballe J (1974) Disjoint Paths in a Network. Networks 4:125-145

104. Thaler DG, Handley M (2000) On the Aggregatability of Multicast Forwarding State. In: Proc. INFOCOM 2000 in Tel Aviv, Israel 1654-1663

105. Thaler D, Talwar M, Vicisano L, Ooms D (2001) IPv4 Automatic Multicast without Explicit Tunnels (AMT). wiki.tools.ietf.org/html/draft-ietf-mboned-auto-multicast-00

106. Thyagarajan AS, Deering SE (1995) Hierarchical Distance-Vector Multicast Routing for the MBone. In: Proc. SIGCOMM '95 in Cambridge, MA 60-66

107. UCLA Computer Science Department (2003) Aggregated Multicast Home Page http://www.cs.ucla.edu/NRL/hpi/AggMC/index.html#_Publications

108. Van Mieghem P, Hooghiemstra G, van der Hofstad R (2001) On the Efficiency of Multicast. IEEE/ACM Trans. on Networking 9:719-732

109. Wall DW (1980) Mechanisms for Broadcast and Selective Broadcast. Ph.D. dissertation, Stanford University

110. Waxman BM (1988) Routing of Multipoint Connections. IEEE J. on Selected Areas in Communications 6:1617-1622

111. Welcher PJ (2001) PIM Dense Mode. http://www.netcraftsmen.net/resources/archived-articles/376-pim-dense-mode.html

112. Williamson B (2000) Developing IP Multicast Networks, Volume 1. Cisco Press, Indianapolis, IN

113. Winter P (1987) Steiner Problem in Networks: A Survey. Networks 17:129-167

114. Yao ACC (1982) On Constructing Minimum Spanning Trees in $k$-dimensional Spaces and Related Problems. SIAM J. Computing 11:721-736